Early Praise for
The Art of Teaching Chemistry

If you are looking for a student-led, inquiry-based teaching approach, you will glean much insight from Mr. Erling Antony's book, *The Art of Teaching of Chemistry*. His teaching technique is one that will challenge many chemistry teachers beyond their comfort level, and it will encourage students to think on their own. This book gives detailed examples on how to broach main topics in a first-year chemistry class, nonstandard labs and demonstrations, and insights on how to become involved beyond the classroom. His passion for his students, the profession, and the scientific community demonstrate his unending love of learning and dedication to the teaching of chemistry. Beginner to veteran chemistry teachers will find this book to be an outstanding guide to improving the level of learning within their classroom of young chemistry scholars.
 -Ms. Angela Koch
 chemistry instructor – Hartland, WI

Mr. Erling Antony has an amazing ability to ignite the natural curiosity of young minds and to demystify the complexities of chemistry for his students. Rather than filling their minds with facts, he awakens their curiosity and teaches them to think. In *The Art of Teaching Chemistry*, Mr. Antony builds on his decades of successful teaching experience in both classroom and lab. He describes teaching strategies that effectively simplify and synthesize the presentation of basic concepts of chemistry in ways that stimulate, engage, and even inspire young learners. This is a great tool for teachers!
 -Ms. Ruth Lynne Walsh
 English instructor – Oconomowoc, WI

I wish this text had been available when I was working my way through high school chemistry. By making students aware from day one of class that they are trustworthy, capable, and expected to have open, questioning minds, Mr. Antony paves the way for students to learn not just facts of chemistry, but also a way of thinking and of approaching all the complexities of life. It is apparent that his expectations of students are very high and that many (or I might guess all) of them exceeded even their own expectations as they rose to those heights.
-Ms. Kathleen Riebau
music instructor (retired) – Durango, CO

I was fortunate to be able to return to school as an older adult and obtain a Bachelor of Science degree in nursing. I had a BA in business, and many of those credits transferred to fulfill requirements for my BSN. However, there was no chemistry to transfer, and the idea of having to take this at a college level filled me with dread. Fortunately, I had a great professor, and all went just fine. I realized, though, just how important chemistry is for so many career paths. A great high school chemistry teacher prepares the student for college chemistry, which can then open the way to careers in health care, engineering and the biosciences, to name just a few. Without this important coursework, these areas will not be reachable. It is a fork in the road.

This is one reason of many that a student should have a great teacher for their first encounter with the subject of chemistry.
-John F. Ward
BA, RN, BSN

THE ART of TEACHING CHEMISTRY

ERLING ANTONY

www.ten16press.com - Waukesha, WI

The Art of Teaching Chemistry
Copyrighted © 2021 Erling Antony
ISBN 9781645383321
First Edition

The Art of Teaching Chemistry
by Erling Antony

All Rights Reserved. Written permission must be secured from the publisher to use or reproduce any part of this book, except for brief quotations in critical reviews or articles.

For information, please contact:

www.ten16press.com
Waukesha, WI

Cover design by Jayden Ellsworth

The author has made every effort to ensure that the information within this book was accurate at the time of publication. The author does not assume and hereby disclaims any liability to any party for any loss, damage, or disruption caused by errors or omissions, whether such errors or omissions result from accident, negligence, or any other cause.

To my loving wife, Patricia, who has endured many evenings and weekends of paperwork.

TABLE OF CONTENTS

Philosophy 9
Introduction 16
Creating a Questioning Mind 22
The Law of Definite Composition 34
The Mole 37
Gas Unit 42
Gay-Lussac's Law of Combining Volumes 46
Formula Writing and Nomenclature 49
Equation Writing and Stoichiometry 52
Stoichiometry 56
Experiments 61
Atomic Theory 66
    J.J. Thomson 71
    Henri Becquerel 72
    Ernest Rutherford 74
    Francis Aston 76
    Robert Milliken 77
    Niels Bohr 80
Electron Configuration and the Periodic Table 85
Bonding, Shapes, and Polarity 98
IMFs and Solution Formation 108
    Solution Formation 111
    Solution Concentrations 114
    Colligative Properties 117

Solution Stoichiometry 121
Acids and Bases 126
Experiments and Demonstrations 133
Additional Experiments 137
    AP Experiments and Other Comments 143
Appendix 1 146
Appendix 2 148
Appendix 3 150
Appendix 4 152
Citations 153
Professionalism/Personal History 155

Philosophy

"The only way to make a man trustworthy is to trust him; and the surest way to make him untrustworthy is to distrust him and show your distrust."
Henry Stinson

Before beginning this text, I feel that I should discuss why I am writing. This text is not meant to be a textbook (that is a book explaining chemistry itself), but rather a text on how to present chemistry to a beginning student. It is certainly the type of text which I could have used as a beginning teacher. I spent many hours in those early years reflecting upon what I wanted the students to know and understand, and how I was going to achieve that. After forty plus years, I still wondered how I was going to better teach some of the topics, but I did discover and create some methods and techniques which were successful in achieving what I had intended. I have tried to include many of those here. I am sure many veteran teachers could add to this list or propose better techniques. If you are a young teacher and practice your craft responsibly, you will soon be able to do the same. Hopefully, some of the techniques and methods within will get you started along this path. I also hope that you will always look forward to the first day of school with anticipation and the last day of the year with some remorse, and that you will start each day with a whistle on your lips.

During my fifty plus years of study in the fields of chemistry and chemical education, both as a student and an instructor, I have observed two basic types of emphasis exhibited by instructors: those who emphasize 'what and how much' and those who emphasize 'why and how.' Of course, neither is exclusive and few if any are one hundred percent in one camp or the other. Most, I would venture, are a sixty-forty split. In fact, while one hundred percent may be possible, if one was totally 'what and how much,' the course would be incredibly boring, little more than a reference book, at least at the introductory level of the subject. Similarly, discussions of 'why and how' have little purpose without some knowledge of 'what and how much.' That being said, there are clearly instructors who focus their attention on the definitions and the mathematics of stoichiometry, equilibrium, and thermodynamics while others focus on the processes. After observing and tutoring students from both camps, and reflecting on my own experiences, I decided that the former group is better at reaching a specific, correct response; that is to say, they are better at standardized exams. However, the latter are more adept at explaining why processes occur and predicting phenomena and results when given a particular scenario. That is, the first group could more accurately determine the yield in a stoichiometry problem, with the correct number of significant figures using standard rules. The latter will give a better explanation of why water has a higher boiling point than the more polar hydrogen fluoride. After reflecting on the issues, I decided to focus on the 'how and why' areas. I do not pretend to know that this was the better choice, but it was my choice. I have taught side-by-side with practitioners of the other camp, and I see and can respect the achievements of their students. I simply want to make this point clear from the beginning. My comments throughout the book are made from the perspective of 'how and why' as opposed to 'what or how much.' After completing the text and receiving reviews from numerous individuals from musicians to nurses, I have received comments ranging from, "I wish my teacher had taken this approach, perhaps it (chemistry) would have made sense," to, "Yes, the logic of my teacher got me through both introductory and organic chem."

As I neared the end of my career as a chemistry educator, I reflected on my evolution as an educator. That change, of course, was most significant during the first few years, but it continued throughout my career. I never believed that improvement wasn't possible. I think that reflection is an important driver of self-improvement. During those early years, I set aside time daily to review basic interactions from the day: Did I treat the students fairly? How could I have done better? If I were a student, would I have felt okay? In short, did I follow the Golden Rule? From the beginning, I emphasized the importance of trust and integrity in the sciences. Politicians and lawyers may not exhibit these traits, but scientists must. We should display integrity as role models, but how do we create trust? Secretary of State Henry Stinson answered this when he addressed Congress regarding the Manhattan Project. Congress wanted to limit the exchange of information between scientists working on various aspects of the project. The scientists argued this was untenable. He told the legislators, "**The only way to make a man trustworthy is to trust him; and the surest way to make him untrustworthy is to distrust him and show your distrust.**"[1]

I had reached the same conclusion years before finding this response. Thus, I shared with my students on day one the fact that I would give them my complete trust. I would not **look** for cheaters, plagiarists and the like. I also made it clear to them that if I discovered that they broke this trust, the penalty would be severe: failure of the class. I shared that if they came to me and said they could not work under those terms, I would quietly help transfer them to a different teacher. Many transfers occurred during the first days of each semester for a wide array of reasons, so the move would not arouse suspicion. In forty plus years, I only had one student ask for a transfer on this basis. Over the years, I had to impose the penalty a few times, but in each case, parents were supportive. The meeting followed a path such as:

I would share my evidence.

Parents would ask the child, "Did you know the consequence?"

Student, "Yes."

Parents, "I guess that is it then. You made a bad decision." Then to me, "How do we proceed from here? What are our options?"

Several students have commented on this trust over the years. The note below, which I received from Ms. Elizabeth Prange, is typical of the sentiment expressed by these students.

> . . . He (Mr. Antony) enacted a policy unbeknownst to most high school teachers called trust. He realized that if he gave his students a large amount of responsibility in their own work, the students would step up with maturity and be accountable for their studies. He impressed upon me the importance of being hon-

est with myself. If I did not finish the homework, yet signed-off saying that I did, I only inhibited my own chance to succeed in this class and in future chemistry courses, but if I were truthful to myself, I took pride in the work I accomplished. Once he showed his trust in me, I returned that trust with a colossal amount of respect.

For all these reasons, he appealed to me both as an educator and as an intellectual. After completing a year of general chemistry and developing an interest in the subject, I decided to take on the challenge of Advanced Placement Chemistry. Knowing that Mr. Antony was the only teacher of AP chemistry at Arrowhead, I looked forward to another rewarding year of expanding my chemistry knowledge.

In his AP Chemistry course, I realized that I shared his same passion for chemistry. Then that passion spread to learning chemistry, and eventually it spread to all my learning. After spending numerous hours a day with him in class, asking questions, and finishing labs, I realized that memorizing equations and facts is not adequate knowledge of any subject. True understanding only begins with this. In order to really appreciate your topic, the theory and explanation of why things happen must be understood. Mr. Antony was the first of my teachers to encourage student-led discussion and stress the importance of asking questions, specifically, the question, 'Why?'

After seeing his excitement when students sought to gain new knowledge, I found myself motivated to try even harder. As new concepts began to click in my brain, I made connections between topics. His enthusiasm increased. He was willing to spend as much time as necessary until the new ideas were solid and cemented into my brain.

After two years of having him as a teacher, it is apparent to me that Mr. Antony has had a positive impact on the way I think. I will be forever indebted to him for introducing me to the wonderful world of chemistry and the vast universe of learning.

Ms. Prange is now a physician.

I am not so naïve as to think that some, perhaps many, students failed to keep the trust, but I know that Stinson's comment is also true for those students whose instructors watch them like a hawk and go to great lengths to catch cheaters. We will not create trustworthy individuals by displaying mistrust. These practices have proven to be successful for my students. When I joined a rural Wisconsin school in the early eighties, thirty percent of our graduates entered college. When I left five years later, over thirty percent of my students entered college pursuing degrees in chemistry, chemical engineering, chemical education or pre-medicine. Giving and receiving trust is instrumental in creating credibility, advising young adults and raising their self-expectations and goals. Most students seem to underestimate their abilities and potential. I recall a conversation with a young lady, Ms. Sue Garwood, in 1977. Sue told me she was going to become a nurse.

I asked, "Why not be a doctor?"

"Oh, I could never do that," she replied.

We talked a while longer. She entered college as a pre-med student. I moved on to a new school the following year and do not know the end of the story, but I had many such conversations over the years, and seldom with a student who set their goals unreasonably high. Perhaps I have been wrong in my approach. Perhaps we need more low skill, low ability, untrustworthy workers, but I don't think that is the case.

Introduction

"That which can be asserted without evidence can be dismissed without evidence."
Christopher Hitchens

Chemistry is often described as 'the study of matter.' Therefore, the study of chemistry matters. It is a topic which should be studied and understood, at least for its methodology, by all citizens. Proper study should lead to a skeptical citizen: one who requires proof, especially of proposals which appear to be contrary to basic principles. For example, claims of alien abduction, interstellar visitation and return. It especially includes claims such as big oil buying the rights to a procedure through which a farm tractor is designed to burn hydrogen, then using some of the energy to complete tasks and using the excess energy to return the water produced, to the elemental state so that the hydrogen and oxygen can be re-burned. This requires 100+ percent efficient processes. Conspiracy theorists make such claims while berating the greed of big business. Big Oil may, in fact, be greedy, but they certainly know and understand the laws of science as well as economics. Individuals with a background in science laugh at such claims, but, unfortunately, many citizens do not have the background and are fooled by false assertion, be they from tabloids or 'news commentators.'

As teachers, we often tell our students that science is not a body of facts but a method of gaining knowledge. Our thoughts, predictions and actions are guided by laws, theories, and principles which have been tested under all conceivable conditions to produce results which are predictable and can be duplicated. Our predictions and actions are not guided by 'gut feel' but by evidence, experience and rational application of principles. As teachers of science, I think it is critical that we have the students understand 'how we know' and help them reach reasonable explanations for the observations. You will find this to be a recurring theme, as I discuss some of the demonstrations, experiments, and class discussions in the pages that follow.

I often introduced the course as one which will rank with their mathematics and physics courses in its demand for exactness in the use of measurement and calculation, but will also challenge their creative side and artistic talents. That is, chemistry involves not only 'what and how much' but also the underlying concepts characterized by 'how and why?' As noted earlier, most chemistry teachers whom I have observed emphasize one of these areas over the other; that is, they focus on the calculations producing students who can perform stoichiometry, equilibrium, and thermodynamic calculations accurately but have little understanding and appreciation for the results, or students who are less adept at the math but understand the underlying principles. Of course, all teachers do both, but most stress one over the other. Personally, I fall into the latter group. This was probably the result of my initial experience with chemistry. I am certain that I caused Mr. Lloyd Haville, my high school chemistry teacher, some difficult times. He would begin class with a call for questions about the lesson of the previous day. My hand always seemed to be in the air asking questions which seemed unrelated to the lesson. Initially he must have thought that these questions were meant to be a distraction. Nonetheless, he would answer each, often with a simple yes or no. But at some point, he began to ask, "Why do you ask that?" I would explain how the principles of that lesson seem to imply X, which combined with Y from the previous unit would lead us to conclude Z. "I don't know if Z is true." If correct, he would acknowledge such. If not, he would patiently explain flaws in my knowledge or rationale, or reveal other pertinent factors, concluding with the actual result. I found the application of atomic level principles to predict macroscopic properties to be all-consuming. I have worked to help my students gain the same passion for understanding. Thus, you will find an emphasis on 'how and why' throughout the text. Chemical purists will shudder at times at the simplistic answers to 'Why?' I do not apologize for this. I believe the first chemistry course should explore basic tenets with cautionary notes of the simplistic answer, with some well-chosen exceptions. The second and subsequent years can focus much more on the exceptions and the mathematics behind the entire process. When exploring exceptions, we should still consider why these exceptions occur, as well as the limitations of our current theory and the student's mathematical ability.

For example, it makes sense to address the apparent inconsistencies of ionization energy from nitrogen to oxygen, that is, going from the P^3 to the P^4 position. But not the electron configurations of palladium which is a D^{10} S^0 as opposed to the D^8S^2 of the other family members. This can be addressed in the second and subsequent courses. Similarly, in the first year we can predict that 2 g of hydrogen gas react with iodine to produce 256 g of hydrogen iodide. We can address the reality of the reverse reaction and equilibrium at a later date after they have mastered the subject of stoichiometry. This does not mean you should mislead the students in the introductory course. On the contrary, you should caution them that we are making some simplifications and the deficiencies will be addressed later in the course or in a later course. I clearly recall a student challenging my comments regarding simplification during the study of electronic structure. Her comment was in effect, "Just tell us the truth." After I wrote the Schroedinger equation for the hydrogen electron on the board, she asked, "What is that?" I responded, "The truth, to the best of our understanding." She requested that we use the simplified version.

So, I will address a number of topics, showing how I approach them and why I chose that approach. I do not mean to suggest that this is the best way nor the only way. I certainly tried various approaches to most lessons: sometimes more than one on a given day if the lesson bombed with the first class. If this occurred, I would revert to an approach which I had used in previous years and which I knew worked but could be improved upon. This is difficult if you are using a PowerPoint, which is part of the reason why I did not use PowerPoints in most of my lessons, except to present pictures or tables and graphs. PowerPoints usually lead to lectures, not class discussions. The latter are more difficult to carry out. But, at least in my experience, students are more engaged and mentally focused during a discussion. Discussion also leads to a less predictable flow of ideas. I often wrote notes for the following day regarding a term or idea which had not arisen during the session but which would be noted at the beginning of the next class.

Twenty years ago, I had a young lady contact me requesting permission to observe a few classes as part of her teacher training program. At the end of the day, she thanked me and said, "I don't believe it. I just saw the same material taught by five different approaches. Virtually everything was included in each, but each was driven by different student questions." The next year, the young lady received permission to student teach with me. The following year, we hired her. Today, she's a highly effective and popular teacher who clearly reflects on what works for her and how to make the class interesting and challenging for her students.

Her observation captures how I liked to run my classroom. I liked to begin with a question, from either the students or myself. After restating or reframing the question, I would ask, "John, how would you respond to Jessica's question?" John usually paused before answering because he knew his answer must include his rationale. If it wasn't included, it would be requested. A response of, "I don't know why," was not accepted. We patiently waited for a response. After John completed his response, we turned to Abby. "Abby, do you agree or disagree with John?" If she agrees, she should add details or examples. If she disagrees, what is the flaw or deficiency in John's response? (*It is important that the flaw is in the argument or logic, not in John.*) Continue the flow of ideas to their logical conclusion. Sometimes this conclusion is false. For example, one day a class reached the conclusion that dihydrogen monoxide should be a high melting molecular solid, like diamond. This was due to a misunderstanding of hydrogen bonds versus bonds involving hydrogen. It was only after someone asked, "Aren't we talking about water?" that they realized an error had been made. We walked back through the logic until we found the false assumption, corrected it and reached a proper conclusion. It is important that you allow some errors to move forward so that they can be discovered and corrected **by the students**. This, too, is good for the mind and the development of logical skills. Of course, some topics, such as nomenclature, cannot be taught this way. But many lessons can be, especially if you are well versed in the subject. This is why I encourage students to earn a major in chemistry if they plan to teach it. Those with only a broad field major often do not have the depth of understanding to adequately address student questions.

Again, this is not a chemistry textbook, nor is it a complete day-to-day discussion of goals, lesson plans, techniques, etc. but is meant to be an aide in helping a young or inexperienced teacher develop lessons which foster understanding and interest.

Creating a Questioning Mind

"The important thing is not to stop questioning. Curiosity has its own reason for existence. One cannot help but be in awe when he contemplates the mysteries of eternity, of life, of the marvelous structures of reality."
Albert Einstein

Before beginning the formal aspects of chemistry, I like to explore a couple of traits of science. This can be set up by asking, 'What are the attributes of a good scientist?' It is easy to generate a list of twenty or so traits which students perceive as attributes for scientists. Some of these, I simply list. Others, like honesty, integrity, good with numbers, etc, I would comment on briefly. Two of these will receive special attention. The first is good observation skills. For years I did the potato demonstration. For those unfamiliar with it, I would have a potato cut to the diameter of a candle. You can do this using a large cork borer to create the fake candle and then wrap in paper-towel until needed, then a pecan or almond is shaved to form a wick and placed in a slot on top of the potato. Divide the class into front and back for the observational competition. Have a student light the wick of the candle in the back of the room while you light the wick of the potato candle. Students make and record observations for 30 seconds. After that time, snuff the flames and compare observations: flame flickers, the flame is yellow, the wick turns black, wax melts and runs down the side of the candle, etc. Invariably, both groups record that last item. When the list is complete, return to this last item. For dramatic effect, bite the candle. The loud snap is impressive, and the students' eyes tell you they are convinced that you are crazy. This is good. Of course, the objective is not to test their observational skills, but to caution them to record what they actually sense, not what they expect to observe. This point can be recalled when investigating in the laboratory or studying the work of Becquerel, Rutherford and others. Proceed with an open mind.

The other traits of a scientist which I elaborate on are good communication skills and teamwork. Written communication, of course, relates to journals, technical reports, etc. To enhance the group cooperation, I divide the class into groups of four. Each group is charged with answering a series of questions, making predictions for subsequent steps and giving supporting rationale. One of the members records and reports for the group. The questions are: 1) I have a large beaker or can being heated on a ring stand above a Bunsen burner. I am going to drop a candle (*yes, this time it is a real candle*) into the beaker. What will happen? Why? 2) I am going to ignite a wood splint and drop it into the can. What will happen? Why? 3) I am going to place 30 mL of water into a small beaker and pour it into the can. What will happen? Why?

Group members must agree on each response. Predictions are reported by each group before carrying out the operations. Typically, the responses are: 1) the wax of the candle will melt, 2) if the flame hits the wick, it may ignite, or else the flame will go out, 3) if the wick was burning, the water will douse the flame. Then we perform the steps. Confidence reigns as the candle melts. It wanes as the splint ignites the contents of the beaker. (*If using a can, you may need to have a tall student don goggles and verify that combustion is occurring in the can.*) Then the eyes bulge with surprise as the water results in a fireball, which will reach the ceiling and extinguish. At this point, they know you are crazy. And they are your disciples. Two obvious reminders. First, of course, goggles and safety materials must be used. They also add to the suspense. Second, a beaker (even a Pyrex beaker) may break under these conditions. Have a fire extinguisher handy. Also, alert your administration that you are doing the activity and that you have tested it when only staff were present, and it poses no danger to students. Additionally, verify that it will not set off fire alarms (*because doing so may provide negative feedback*). Then when the eyes return to normal size and breath returns to their lips, share a line such as, "I contend that any of you who have watched a candle burn should have known what was going to happen before we carried out these steps. Your assignment for tomorrow is to go home, help prepare a nice dinner, light a few candles, and discuss with your family how a candle works. Be ready to discuss it tomorrow. Yes, I know the answer is on the web, but do yourself a favor and don't look it up. You can figure this out."

    The following day, we devote the first part of the class to the sequence. It is important to address the question, "How did we know that wax burns?" Thought experiment. You say, "Suppose you have two identical strings and ignite them. How does the amount of heat and light, that is, energy, released compare?"

"It should be the same."

You, "Good. Now use one of the strings as the wick of a candle. Burn the strings. How does the heat and light produced by the bare string compare to that of the candle?"

"The candle has much more."

You, "Right. Why?"

"Either the wax burns and produces heat and light, or the melting and vaporizing of wax produces the heat and light."

You, "Right, so when you melt something, do you put heat into the object, or is heat released from the object?"

"It is absorbed."

You, "Right again. And when you vaporize a liquid?"

"You have to add heat as well."

You, "Right, so if the paraffin absorbs heat and light as it melts and as it vaporizes, should we experience more or less heat released from the candle compared to the bare string?"

"Less."

You, "Yes, but we experience more, so . . ."

"Wax burns."

As an afterthought, "When do we smell a normal candle?"

When the flame is snuffed."

You, "Yes. Why?"

You have also set the stage for two recurring themes. The first is the opportunity to make them think about how nature works, and the second is to use their current knowledge in new ways.

An exchange like this is important because it will cause them to begin thinking about how and why things happen. This should be reinforced frequently throughout the course, as should applications of phenomena, such as the use of neon lights and the study of Red Shift during the electron transitions lesson, or the use of limestone in streams to neutralize acid rain while creating safe havens for trout and other game fish. They challenge the students to begin asking themselves, "How does that happen?" or, "Why does that happen?" as they did as three-year-olds, but now they can answer many of these questions using knowledge and experiences which they have already gained.

The other topics which I emphasized during the preview unit are mathematical in nature. First, the use of significant figures. It is important that we emphasize the link to measurements. To ask, "How many significant figures does the value 6.10 contain?" is ludicrous without asking, "What is being measured?" If the length of a benchtop is found using a meter stick with 0.05 cm scaling, then there are three significant figures, but if it is a pH reading then it only has two significant figures (as the 6 is just a decimal point placeholder): the 6 is just a number, not a significant figure. The term 'significant figure' has no real meaning unless related to a measurement or calculation based upon a measurement. Similarly, the operational rules should be explained. For example, suppose a trucker arrives at a weigh station then goes into the office area while his truck is being checked. When he returns, he receives word that the operating gross weight is 64,500 pounds. This is plus or minus 100 pounds. *Not really,* he thinks, *because I weigh 212 pounds plus or minus 1 pound and I bought this soda which is 12 ounces or .75 pounds within .01 pounds. And a fly also came into the truck. That would be an additional .0001 pounds, so the total weight is 64,712.7501 pounds.* How much of this weight can be considered reliable? Clearly, the fact that we rounded the truck's weight to within 100 pounds means that the soda and fly had no measurable effect and would be undetected by the truck scale. In fact, the trucker himself is barely detectable.

You have set the stage. Now you can present the rules and have them make sense. If asked, you can make up a problem to demonstrate the other rules. For example, suppose you are asked to demonstrate the rules for multiplication and division. You can find the area of a benchtop. If measurements show it to be 65.3 cm by 24.7 cm, what is the area? Multiplication shows the result to be 1612.91 cm^2. How many of those digits can you trust, including the first digit of uncertainty? Given the uncertainty in each measurement of perhaps .2 cm, what is the largest and smallest possible area? Show the calculations and how the results agree with the rules for significant figures. Students believe when they see a reason. That does not imply that they can or will immediately apply the rules religiously, but at least they are not applying the rules and asking, "Why are we doing this?" Caution. You should also warn the students that the rules they learn only work some of the time. The *Journal of Chemical Education*[2,3] has had several articles on the use and limitations of significant figures. I suggest you access these if you have questions. I also refer the students to the use of a better technique, Differential Error Analysis, as a method which they will learn when their math skills have improved.

The other topic which I emphasize during the preview is graphing. I include this under the umbrella of "How do we know?" A large percentage of the equations we encounter in science fit the format: $Y = m X^a + b$ where 'a' is 1, ½, 2 or negative 1, or occasionally other similar values, such as 1/3, negative 2, 3 etc. I also note that there is another common relationship, that being exponential growth, where $Y = m\, a^x + b$, but most sophomores are not familiar with exponential and logarithmic growth, so I do not include the analysis of these relationships at this time. A similar approach can be applied after they gain the mathematical foundation.

Graphing

We begin this fundamental unit by reviewing what they learned in algebra and other math classes since then, especially the basic equation of a line, slope and intercept, emphasizing that **all** straight lines and **only** straight lines fit this equation. We also discuss the physical significance of slope and intercept for a graph of data. This includes a short assignment where we plot some inane data and interpret 'slope' and 'intercepts.' The next day, we dive into non-linear graphs. Students choose values for 'm' and 'b', calculate several X/Y pairs and graph them to see the shape. We then examine the shapes produced. That is, is it parabolic, hyperbolic, etc, and how does each relate to a particular value of the 'a.' Most accept this quite readily. So then I ask, "Suppose we are in the laboratory and collect data and then plot the data. Can we go to the graph and determine the general equation?" Yes. If it is parabolic in the first quadrant, it implies 'a' is 2 [or perhaps 3, 4, etc.). But how can we be sure that 'a' is 2, and how do we know the values of 'm' and 'b'? Answer: we return to the general equation $Y = m\, X^a + b$ and define a new variable, perhaps Z, as $Z = X^2$. This produces an equation of the form $Y = mZ + b$ which will yield a linear graph iff 'a' is '2.'

Let's see how this works. Suppose we generate the following data:

X	Y
1	5
2	14
3	29
4	50

Plotting the points gives a parabolic shape. So we define Z as X^2 and determine the values of Z as below.

X	Y	Z
1	5	1
2	14	4
3	29	9
4	50	16

Now plotting Y vs. Z yields a line of equation $Y = 3 Z^1 + 2$, but since Z is X^2 we have $Y = 3 X^2 + 2$ as the equation which fits the initial data. Try it.

But suppose the actual equation had been $Y = 3 X^3 + 2$. Then the data and calculated Z would have been:

X	Y	Z
1	5	1
2	26	4
3	83	9
4	119	16
5	377	25

The plot of Y vs. Z would still be a curve, but if we now define Q as X^3 then we have:

X	Y	Q
1	5	1
2	26	8
3	83	27
4	119	64
5	377	125

The resulting plot of Y vs. Q is linear with Y = 3 Q + 2.

We then do a simple lab measuring things like the mass vs. the length of a side for paper squares, distance vs. mass needed to lift a given mass using a first class lever, etc. This gives students some confidence in the processing of data before using it in the chemical laboratory.

As you work through this unit, be certain students realize the significance of 'm' and 'b' for all linear graphs and, when reasonable, nonlinear relationships. For example, if you place varying amounts of a liquid in identical cylinders then measure the volume and mass, graphing the mass vs. the volume of liquid, what do 'm' and 'b' represent? (*The density and mass of the empty cylinder.*) Similarly, if you determine the mass and length of a side for different squares of a given type of paper, a graph of mass vs. length gives a parabolic plot. But a plot of mass vs. length of a side squared is linear. So the slope is the mass of one square centimeter of the paper. The significance of some of the 'm' and 'b' values will be difficult to identify for nonlinear graphs and can be ignored for the present time. This is especially true of inverse relationships, such as one receives when determining the mass needed to lift a given object vs. the distance from the fulcrum. Later, they will recognize this as torque but probably won't have this knowledge at this point. These are simple measurements, but they also teach about scatter and reinforce the need for significant figures.

Various software companies offer software programs which allow the students to try different values of 'a' very quickly and, once a linear relationship is found, determine the values for 'm and b'. I prefer this approach rather than using the solver on a calculator because 1) Mother Nature seldom uses an exponent such as 3.274 or a seventh-degree polynomial although the calculator gives such as the best fit to the data and 2) students understand what they are doing and why it works. The calculator can be just another black box providing an answer without understanding. This technique allows the student to manipulate the data into an equation in a logical manner. Of course, it will not work if the equation is of the type '$Y = 2X^2 + 3X - 4$'.

These are the three topics I consider key to the preparation of the student mind for chemistry. 1) Creating a questioning mind and requiring a rational explanation to support conceptual answers. 2) Use of measurement and understanding of the limitations of these measurements. 3) Use of graphs and fundamental algebra to determine mathematical relationships.

Other topics are certainly important, such as the use of the metric system, but I have some specific thoughts when teaching those materials. With respect to measurements, I think it is important that you also discuss relative vs. absolute scales. Most of our scales are absolute: 0 feet, 0 dollars, 0 pounds, 0 seconds, etc. Some, however, are relative in that 0 is not the minimum possible value. There is no time shorter than 0 seconds nor distance less than 0 feet. However, there are temperatures less than 0 F or 0 C but not 0 K or 0 R. It is important that the student sees the difference, and it will help during the gas unit if they understand the difference. The Pauling Electronegativity Scale is another relative scale. I think it is also important that they realize that all of the scales are, or at least were, arbitrary in their definition. A yard could have been defined as 4 feet, which in turn could have been defined as the distance between the big toes when Henry VIII stood heal-to-heal. The definition was arbitrary, but we must use it as defined. I also have an opinion on Dimensional Analysis. Tutoring students from other teachers, especially at the collegiate level, I have found that many (perhaps most) students don't understand the process. They become adept at using it but don't understand the process conceptually. I believe it should be taught as a variation of the Multiplicative Identity Property in mathematics. That is A×1 equals A, and 4×1 = 4 equals 3+ 1 equals 2*2 equals 16/4, etc., such that any ratio of these expressions such as 2×2/(16/4) equals one: that is, unity. Similarly, a ratio of identical quantities also defines the value one. Since 36 inches equals 3 feet or 1 yard, a ratio of any two gives unity. Thus, we can use the Multiplicative Identity Property to convert units.

For example, 180 inches = ___ ft can be solved as ___ feet = 180 inches = 180 inches * 1 = 180 inches * 1 foot/12 inches.

The unit inch cancels, giving an answer of 15 feet.

If they understand this concept, they can work through the subtle changes which occur when we use ratios for chemical change, such as 2 O_3 <-> 3 O_2 which yields 2 moles O_3/ 3 moles O_2

Personally, I did not teach Dimensional Analysis the last twenty-five years of my career, choosing instead to use proportions. Conceptually, students understand that 2 moles of ozone produce 3 moles of diatomic oxygen, therefore 20 moles of ozone should produce 30 moles of diatomic oxygen much better than they understand:

___ moles O_2 = 20 moles O_3 * 3 moles O_2/ 2 moles O_3

Proportions are more logical and conceptual than dimensional analysis. As a result, my students might have to go through several steps such as converting grams of A to moles of A then moles A to moles B and finally moles B to liters B. Four steps, but they understand each step. After gaining this understanding, they can apply the methods of Dimensional Analysis, which they learn in physics, if they wish. I prefer thorough understanding of an answer more than a quick or simple determination of the answer.

The Law of Definite Composition

Never trust atoms: they make up everything.

I feel that many texts give little attention to a very fundamental concept with great implications. This is the Law of Definite Composition. I tell the story (*essentially a historical fiction but very useful as a teaching tool – whenever I do this in the course, I make the students aware that the principles are solid, but I have embellished the details*) of how scientists moved chemistry from the practice of alchemy to a true science. This happened when its practitioners began to meticulously measure and record the quantities of materials consumed and produced in reactions. John Dalton and others completed much of this work. Thus, for example, they produced mass ratios such as:

H + F → HF
1.0 g + 19.0 g → 20.0 g

H + Cl → HCl
1.0 g + 35.5 g → 36.5 g

H + Br → HBr
1.0 g + 79.9 g → 80.9 g

H + O → HO
1.0 g + 8.0 g → 9.0 g

Fe + O → FeO
7.0 g + 2.0 g → 9.0 g

Furthermore, when compounds were decomposed, the same ratios were observed. Seeing definite ratios such as this allowed the scientists to predict other ratios, such as 28.0 g of iron should react with 35.5 g of chlorine. When these ratios worked, the scientists were able to logically formulate a set of principles regarding atoms. 1) When elements react, they do so in definite whole number ratios to form a compound. If this were not true, sometimes 1.0 g of hydrogen would combine with 8.0 g oxygen, and the next time 1.0 g hydrogen might combine with 7.0 g or 9.0 g of oxygen. This was not observed. 2) All atoms of an element must be identical. If some oxygen atoms have a mass of 8.0 relative to hydrogen and others have a mass of 10.0 or 5.0, then different samples of water should yield different mass ratios for the elements depending on which mix of oxygen types is present. 3) Since we can combine certain substances to form new products which can then be returned to their original state but not further, the first type must be the fundamental building blocks of nature (i.e. elements). The atoms of these elements cannot be created nor destroyed. That is, they are indestructible and eternal. Thus, Dalton envisioned atoms which are like billiard balls: solid spherical bodies which are identical for a given element but vary in size and mass by element. Thus, a hydrogen atom might be likened to a BB, while an iron atom is comparable to a ball bearing, and uranium is like a bowling ball (without the finger holes).

This also allowed the atoms to be arranged by relative weight. Hydrogen, always seeming to be the lightest, could be given a weight of 1.0, causing oxygen to be 8.0, fluorine 19.0, chlorine 35.5, and so forth. Note that at this time, water was assumed to be HO. Later, the discovery of diatomic elements and oxidation states led to corrections in these values and assisting Mendeleyev and others with the development of the Periodic Table and the Law of Multiple Proportions.

The Mole

"Nothing in life is to be feared, it is only to be understood. Now is the time to understand more so that we may fear less."
Marie Curie

When I started teaching, I incorporated the concept of the mole within other units as it became needed. Later, I learned to make it a separate one-day unit with noticeable hype and hoopla. The weekly schedule would contain large, colorful print with notes such as DON'T MISS THIS DAY!! Coupled with the approach below, most students picked up the concept readily and could use it in their logic quickly. Understanding the particulate character of the mole is key to applying it when considering atomic weights, stoichiometry and many other fundamental concepts. This approach also aligns with my approach to stoichiometry as you will observe on page 57. I begin with units which are familiar to the student and progress to less familiar units of measurement. It proceeds something like this:

Suppose you are the manufacturer of balls used in games and athletics. You produce ping-pong balls which weigh 1 ounce each, golf balls weighing 4 ounces each, tennis balls weighing 7 ounces each, softballs at 10 ounces each, and bowling balls at 238 ounces each, as well as various other balls. (*Display or describe vertically on the board or on a projected screen.*) Stores can purchase the balls in boxes of sixteen. What would a box of the ping-pong balls weigh? 16 ounces, or 1 pound. What about a box of the golf balls? 4 pounds. The tennis balls? And so forth. (*Show results on the display as you continue.*) Why does this work? Well, because the weight of the box of balls is sixteen times the weight of one ball and then, to convert to pounds, we divide by sixteen, which is the number of ounces in a pound. Thus, the unit changes, but the numerical value remains the same. (*This is a place where dimensional analysis may be helpful.*) Likewise, if you sell cases of 2,000 boxes, a case of ping-pong balls would weigh 2,000 pounds or one ton, a case or 2,000 boxes of tennis balls would be four tons and so forth. Again, we see the weight being: the weight of one box in pounds times 2,000 boxes divided by 2,000 boxes per case. We could do a similar calculation relating the weight of one ball to the weight of a case if we realize there are 32000 balls in a case and 32000 ounces in one ton.

Would this work with the metric system? Of course it would, but the multipliers would be different, becoming 1,000 if converting from milligrams to grams or 1,000,000 if converting milligrams to kilograms. Should it work for any other defined unit? Of course, as long as the multiplier and divisor are the same value, the unit should change while the numerical value remains the same.

As chemists, we saw in the introductory unit that Dalton and others developed the Law of Definite Composition, which implied that each element has a relative characteristic weight. So, if we assign the lightest element, hydrogen, a mass of 1, helium has a relative mass of 4, lithium a relative mass of 7, and uranium a relative value of 238. That is, 238 hydrogen atoms have the mass of one uranium atom, etc. Then some number of hydrogen atoms should have a mass of 1 gram. We'll call this number X. This same number, X, of helium atoms should have a mass of 4 grams and X atoms of lithium have a mass of 7 grams, while X atoms of uranium are 238 grams. Just as 32,000 changed the weight of the ball in ounces to that numerical weight in tons, X changes the mass of a single atom in atomic mass units to that same numeric value in grams. Thus, X atoms of nickel have a mass of 58.70 grams and X atoms of . . . (*pull random elements from your Periodic Table*). Check for understanding and take questions, then proceed. What would be the mass of 2X atoms of nickel?

Finally, it is time to give a name to X. This number, X, is what the chemists call a mole. The most similar unit in our language is a dozen. A dozen always means 12. A dozen eggs are 12 eggs, a dozen donuts are 12 donuts. A dozen workers are 12 workers. In the same way, a mole of eggs are X eggs, a mole of donuts are X donuts, etc. The only thing which makes a mole special is that it converts the relative mass of an atom to that same number of grams of atoms. Thus, a mole of carbon atoms has a mass of 12 grams, while a mole of uranium is 238 grams. The same number of atoms of uranium have a mass nearly 20 times greater than that number of carbon atoms because each uranium atom is nearly 20 times heavier than a carbon atom. Likewise, a mole of molecules would have a mass in grams of the sum of the masses of the individual atoms involved. Thus, Li_2O would have a mass of 30 grams per mole (7+7+16). This quantity is termed the molar mass, or sometimes the atomic or molecular weight.

So the only question remaining is, "What is the value of X?" "Is it large or small?" I liked to take suggestions. Then write 602 and then let them begin to add zeroes. 602,000? No, more. 602,000,000? More. They usually quit after about 602,000,000,000,000. Keep adding zeroes to reach Avogadro's Number, and then shorten to 6.02×10^{23} (reminding them of scientific notation). Label the term and emphasize that the number is not as important as the fact that such a number **must** exist and that it is important that we know the actual value. The idea of the number was created long before its value was known and made possible the pioneering work in stoichiometry and atomic theory. (*Be ready to answer the question, "How do we know, or how did we measure this value?"*)

To allow time to absorb this key idea, I explored the magnitude of Avogadro's Number by using the long expression. Starting at the units digit, I work to the left giving familiar quantities: the number of people in the room over age thirty, the number of students in the room, the number of juniors in the school, the number of students in the school, the number of citizens in the community, number in the county, the number of people in the state, the number of people in the country, the number of people in the world, the number of people who ever lived. Hmm, have to change units. Size of the US debt in dollars. Size of the debt in pennies. And still, we have a long way to go. This should give students an appreciation for the magnitude we are talking about. This is approximately the number of molecules of water found in a tablespoon of water. The *Journal of Chemical Education* (referred to hereafter as *JCE*) has published several articles over the years which are helpful to student understanding of this value.[4,5,6]

By now the class period will be nearly gone, and you may wish to generalize the equation for calculating mass or moles

(# moles = # grams/molar mass) and then give them a few problems. This approach allowed me to focus on the atom as a particle, noting that particles combine or interact in a definite and consistent number ratio. Thus, two hydrogen atoms can combine with one oxygen atom or with one sulfur atom, although the mass ratio for the two substances are 1 gram hydrogen to 8 g of oxygen or 1 gram of hydrogen to 16 g of sulfur.

[a] Chemists of the world, I realize the mole is not just a number, but I prefer to focus on the large concept first and then clean up the details later.[7]

Gas Unit

"Science is not competitive but cooperative."
Linus Pauling

After establishing the Law of Definite Composition and the mole, it is time to examine gases. As noted during the introductory unit, the gas laws are an obvious application of graphing techniques. The gas laws can be easily developed using experimental data and graphs, including pressure vs. volume, pressure vs. temperature, volume vs. temperature, pressure vs. moles, and volume vs. moles. You may have the entire class do each experiment or divide the class into smaller groups and have groups explore the relationships. I recommend having several groups collect data for the volume vs. moles of reactant and, by extension, gas produced.

This is a nice place to allow students to design the basic experiments, and you can expand or correct as needed. The volume vs. moles can be easily determined using reactions such as magnesium or zinc with excess hydrochloric acid or baking soda with vinegar. After collecting data, graph the results and determine each equation, including interpretation of the physical significance of slope and X and Y intercepts. The groups should then share the findings with the rest of the class. You should explore things like the importance of the X intercept in the pressure vs. temperature graph. Why? And the pressure vs. the inverse of volume graph (which yields a linear graph). I used this as the first of numerous thought experiments. We have just finished changing the volume of gas in a syringe to the inverse of volume. A sample exchange might proceed as:

You, "So we have just finished plotting P vs. 1/V to obtain a straight line. The Y intercept is __ (*some number close to 0*). When would 1/V equal zero?"

Student, "It can't. 1/0 is undefined."

You, "True, but we are not asking 'What is 1/0?' but rather, 'When does 1/V = 0?'"

Student, "It can never be 0."

Student #2, "But if a very big number is used for V, it becomes very close to zero."

You, "Right, so when does 1/V become extremely close to zero?"

Student, "When the volume is very large."

You, "Okay, so what is the largest volume you can think of?"

Student, "The earth? No, the solar system? No, the universe."

You, "Okay, so thought experiment. Suppose everything in the universe is gone except for this syringe and the gas within it. Now the gas is freed and allowed to spread for a kajillion years so that the gas occupies the volume, which is our current universe. Then we pick a random spot in the universe, say the point where the tip of this pen is currently located, and measure the pressure at that point. What would be the pressure?"

Student, "The pressure would be zero."

You, "Good. So now the equation changes from P = #/V + b to P = #/V or PV = #. This equation should work for all points on the line so, if we choose two random points 1 and 2, $P_1V_1 = \#$ and $P_2V_2 = \#$ so $P_1V_1 = P_2V_2$, Boyle's Law."

Then you should generalize their equations, showing how the equation can be easily modified to give the other gas laws: Charles's, etc.

Vernier has some nice, interfaced experiments for the first several experiments.

After deriving these equations and noting their restrictions ($P_1V_1 = P_2V_2$ if temperature is constant and there are no leaks [n is constant]), a bit of algebra gives you the combined gas law, $P_1V_1/n_1T_1 = P_2V_2/n_2T_2$ and then the Ideal Gas Law. Another little lab gives you the value of the Universal Gas Constant, R. This is nice because it is all laboratory-based and easily completed in a week, even with forty-minute lab periods as we had.

Students should be able to list six properties of gases. Be certain to distinguish between properties of some gases and those of all gases. Once a complete list is generated, you may ask questions such as, "How do gases create pressure?", "Why does pressure increase with temperature at constant volume?", "If gases are compressible, but liquids and are not, what does this imply about the molecules of each?", "Are there other properties that suggest these characteristics?" These points generate the Kinetic Molecular Theory for gases. Each point in the theory should be supported by one or more observed properties. Emphasis here, of course, is on the particulate nature of gases.

Students have an intuitive sense and can visualize the molecules moving faster as the temperature rises. This causes more and 'harder' collisions with the walls, which increases volume or pressure or both. Likewise, condensation takes place when the attraction between between the particles overcomes their kinetic energy. (*A preview of IMFs. Why does the heavy HI molecule have a lower boiling point than water molecules, although they travel at only one fourth the speed?*) Present this concept, as well as longer-term questions like, "Why do some gas collisions result in explosive change while others are perfectly elastic?" (*This is a preview of collision theory.*)? As noted earlier, I enjoy these 'why' questions which give the students something to think about and puzzle over. Also, don't become so focused that you do not return to the properties of some gases: "Why do some have an odor while others do not? Why are some combustible while others are not?" and so forth. You need not answer all or any of these but use them as food for thought. Don't be surprised if you receive a question such as, "So is that why some gases burn?" when studying enthalpy and bond energies.

Given this, they are now ready to derive the Law of Combining Volumes. This discussion should include the discovery of diatomic elements, gas phase stoichiometry and the basics of chemical equation writing. See below for one approach.

By the way, gases can be a nice springboard into content. That was the historical basis for much of chemistry including stoichiometry, the particulate nature of matter, the mole, and other topics.

Gay-Lussac's Law of Combining Volumes

"Everything must be made as simple as possible. But not simpler."
Albert Einstein

This law is the key to gas stoichiometry, but I use it in a more historical manner, taking some liberties, of course. Earlier in the year, we developed the Law of Constant Composition and the idea that if 1.0 gram of hydrogen combines with 8.0 grams of oxygen, and 1.0 gram of hydrogen combines with 23.0 grams of sodium, then we should expect 8.0 grams of oxygen to combine with 23.0 grams of sodium if they react with each other. They do so. Evidence of this type led to the formulation of that law.

I developed a parallel approach to the Law of Combining Volumes. After describing the objective for the study (**to determine the ideal ratio for the volumes of reacting gases at constant temperature and pressure**), I would ask for designs which allowed us to collect such data. Students could quickly describe a system which would allow one to measure the volume of gas one, then of gas two at the same T and P, mix the two gases, spark the mixture, allow to cool and adjust to the original pressure, and then check for excess reactant. Most groups chose a piston fitted to a cylinder to allow a change in pressure which would self-correct to the original pressure. After you get the mixture just right, complete the reaction again and measure the volumes of product gases. Doing so would result in values such as these (*this is not Gay-Lussac's data*). All reactants and products are in the gaseous state.

H + F → HF
1.0 L + 1.0 L → 2.0 L

H + Cl → HCl
1.0 L + 1.0 L → 2.0 L

H + Br → HBr
1.0 L + 1.0 L → 2.0 L

H + O → HO
2.0 L + 1.0 L → 2.0 L

H + N → HN
3.0 L + 1.0 L → 2.0 L

N + O → NO
1.0 L + 2.0 L → 1.0 L

In all cases, Gay-Lussac found that the reactant and product gases formed small, whole number ratios, i.e. the Law of Combining Volumes. But why? (*I've been told, although I have never seen it verified, that Gay-Lussac speculated that equal volumes of gas at the same temperature and pressure may contain equal numbers of gas particles regardless of size, density, complexity, etc.)* Examination of the data, however, denied this possibility: if one atom of hydrogen combined with one atom of chlorine, one molecule of hydrogen chloride would form. If his speculation was correct, one liter of each reactant should result in one liter of product, not two. His student, Amedeo Avogadro, solved the puzzle by proposing diatomic elements. He reasoned that, if hydrogen and chlorine are both diatomic, the atoms could separate and produce two molecules of HCl, agreeing with the experimental results. This also implied that fluorine, bromine, oxygen and nitrogen were diatomic, and that water was H_2O, not HO as previously assumed. This, in turn, implied that the mass of oxygen relative to hydrogen was 16, not 8. Other elements which had a relative mass based upon oxygen also needed correcting. Thus, we can see that Gay-Lussac and Avogadro laid the foundation for gas phase stoichiometry, a method for determining the formula of a gaseous compound, the balancing of gas phase chemical reactions and the concept of the mole. It was appropriate that, when the number of particles required to describe a mole was determined, the quantity was termed Avogadro's Number.

As is frequently the case in science, we assume the simplest possible relationship until we find contrary evidence (HO vs. H_2O). Then we again develop an explanation which is as simple as possible while including all evidence. We see this with Avogadro. He could have proposed that chlorine was a tetratomic or hexatomic and the products were HCl_2 or HCl_3 but there was no evidence requiring this degree of complexity.

Formula Writing and Nomenclature

"Never memorize something you can look up."
Albert Einstein

I have no great insights into teaching this extremely important unit except, students MUST learn the common ions (including their charges) or you will lose hours to the problem over the course of the year. Once they know that potassium is K^{1+} and sulfate is SO_4^{2-} writing K_2SO_4 and naming it potassium sulfate becomes easy. Not knowing these ions makes the task very tedious and difficult. Typically, I am not an advocate for memorization nor drill and practice, but here it pays off. The same is true for equation writing. Quiz them repeatedly on the ions. Since I did not correct nor grade homework, I provided plenty of worksheets with compounds to write or name on the front side and answers on the reverse side. Students could practice as needed to gain mastery.

This being said, you can help the students by pointing out a few keys, such as polyatomic ions containing oxygen usually contain the name of the central atom as a base followed by the suffices 'ate' or 'ite' with the 'ate' ion containing one more oxygen than the 'ite' ion, as in nitrate versus nitrite, sulfate versus sulfite, and chlorate versus chlorite, but both ions carrying the same charge. The even or odd charge of the polyatomic ion is usually the same as that of the non-oxygen, central atom (nitrogen, phosphorous, chlorine, bromine, etc. are odd while carbon, sulfur, silicon, etc. are even). Addition of a hydrogen atom adds one positive charge (PO_4^{3-}, HPO_4^{2-}, $H_2PO_4^{1-}$). Of course, if you have already studied bonding, you may show them the common bonding pattern, and the charge need not be memorized as it could be generated if uncertain. This would reduce the amount of brute force memorization, but I would still recommend memorization.

After mastering nomenclature, I usually took a few minutes to discuss the 'old way' of naming multivalent cations, warning them that, when in the laboratory, they will probably not find a bottle of iron (III) chloride or copper (II) sulfate but instead would find ferric chloride and cupric sulfate. Then explain that that the 'ic' suffix implied the higher of two common valences while 'ous' implies the lower charge. This is a good opportunity to include some history. Pb^{2+} is plumbous because the Latin name for lead was plumbum and is the base for our terms plumber and plumb bob. Similarly, Fe was the symbol for ferrum, thus ferrous and ferric ions. Some metals use symbols based on early names but use English names so that Hg_2^{2+} and Hg^{2+} are mercurous and mercuric, not the Greek hydrargyrous or hydrargyric.

If you are a punster, you can have fun here. Fortunately for salad-eaters we use the term plumbous not leadous, and Fe^{3+} is ferric not ironic. Co^{3+} was discovered in a bay adjoining the Baltic Sea, the Cobaltic Sea, and, of course, the most popular of all ions is St. Ni^{2+}. Don't forget the ferrous wheel.

In conclusion, if students have a firm grasp of the common ions, they will master nomenclature and formula writing. With this they will have a much easier time with equation writing and stoichiometry. If they do not gain mastery of the ions, their trouble is just beginning. I would suggest the inclusion of an exploratory experiment during this unit. I called mine the Ionics Vs. Covalents lab. The simple experiment consisted of six or seven stations where students compared 'random' ionic and covalent compounds for 1) aqueous solubility, 2) nonpolar solubility, 3) conductivity as an aqueous solution, 4) conductivity of a melt, 5) odor, 6) crystallinity and 7) flammability. This simple experiment provides significant insight into the two types of compounds and is frequently referenced in later class discussions.

Equation Writing and Stoichiometry

"If you are not part of the solution,
you are part of the precipitate."

After completing formula writing and nomenclature, equation writing and balancing equations is a logical next step. I have no great insights except that, if the student has not mastered formula writing, they should be pushed to do so before continuing. The time they save and the frustration and headaches they prevent will justify the expenditure of time; encourage them to do so. Students will find this unit **much** easier if they have memorized the common ions and their charges. Allow me to repeat that; students will find this unit **much** easier if they have memorized the common ions and their charges.

There are numerous approaches suggested by textbooks and all work if the student understands a few requirements. First, as with formula writing, all formulas must be neutral. BaCl is not written correctly nor is HO nor NaS. Here is where that mastery of ions pays dividends. Second, polyatomic ions remain intact except for specific ions which may decompose under strong heat or as specific compounds. Third, oxidation states do not change during double replacement (ionic) reactions. They will change during single replacement and synthesis reactions, and they may change during a decomposition reaction.

I found the students had the most difficulty with single replacement and decomposition reactions, so I began with the easiest (synthesis) and proceeded to double replacement reactions. This allowed them to battle the subscript vs. coefficient fight with which many struggle. By the time they work through several sheets of problems, they have mastered that difficulty. It also allows some lab time with numerous reactions which require balancing before taking on the other two types of reactions.

There are many sources of common lab experiments which demonstrate the various types of reactions, and I have little to add to the synthesis, decomposition and single replacement experiments except to say, for the latter, choose a series of reagents which clearly show an activity series so that you may relate to the table of activities. I did create a spin on the double replacement reaction which I found useful (Appendix 2). I set up twenty-five to thirty different lab stations around the room, including a couple in each fume hood. Students started at some random station near their personal lab station and worked their way around the room mixing a couple drops of solution one with a couple of drops of solution two, making observations of heat produced, solid produced or odor produced (*do those in a hood*). They then clean their test tube and continue to the next station (*note that if a toxic metal is used, the waste and wash should be caught in a beaker for later disposal*). The stations were designed to randomly illustrate the products which are gases or in which the product decomposes to form a gas, products which release detectable heat, products which include a precipitate and mixtures which show no noticeable change (no reaction). After completing their work, students are asked to identify which of the products was responsible for the observed change, if there was a change. For example, products at station two might be $BaSO_4$ + KCl and students observe that a white precipitate forms. Which product is responsible? If you look at the products from station one, they may be $NaNO_3$ + KCl and no precipitate forms. Since KCl was common to both sets of products, it must not be the precipitate which implies that the $BaSO_4$ was responsible. Furthermore, if we look down to station eleven, $BaSO_4$ is again a product, and we again see a white precipitate along with a blue solution containing $CuCl_2$. After completing this analysis, it is easy to pull out the products responsible for gas formation and heat production, and the solubility table which you supply makes sense. If the substance is not soluble in water, it becomes a precipitate while those salts which are soluble remain in solution. Thus, you can ask them to predict the result of various mixtures. Do so, asking them what they would like you to mix and predict the result before actually mixing the solutions. Point out repeatedly that these are solutions of the ionic salts, not the salts themselves. They know this in the back of their minds, as they have done an experiment comparing the properties of ionic and covalent compounds and know ionics to be high melting, crystalline solids, but this reminder will still help drive that fact home. Try several combinations, noting that you can predict the formation of a solid, but not the color (ammonium sulfide with cobalt II chloride for example).

Before leaving this topic, the students should be able to recognize the type of reaction based upon the reactants, predict the product(s) and balance each type of reaction. They should also be able to predict if a single replacement is likely to take place and the results of a double replacement reaction (heat, gas, precipitate, no reaction). You may choose to include combustion reactions here.

I did not address redox reactions at this time, choosing to study them later in the year after students are more adept and confident and can handle the use of ions in net reactions. From here, we proceed to stoichiometry.

Stoichiometry

"Chemistry itself knows altogether too well that – given the real fear that the scarcity of global resources and energy might threaten the unity of mankind – chemistry is in a position to make a contribution towards securing a true peace on earth."
                    Kenichi Fukui

As we enter this unit, students should be familiar with balancing equations and the concept of the mole. If these are understood, the rest is easy.

I like to use a synthesis reaction as an example, one which involves three different coefficients so that students may easily sort them out. For example, $2\,Al + 3\,S \rightarrow Al_2S_3$

What does this mean? Well, it is like a recipe for making aluminum sulfide. Note that the reactants are elements, so each is neutral. The product, of course, is also neutral but consists of aluminum (+3) ions and sulfide (-2) ions. Thus, during the reaction, the → portion of the statement, each of the two aluminum atoms lost three electrons (recall from previous classes that electrons are negative), and each of the three sulfur atoms gained two electrons. (Draw pictures to illustrate this point.) (*Practicing chemists please note, this is not meant to be a pro-*

posed mechanism for the reaction but just a simplistic approach to help the student interpret the balanced equation.)

The Xs in the diagrams represent electrons.

First, sulfur atom #1 takes 2 electrons from aluminum atom #1.

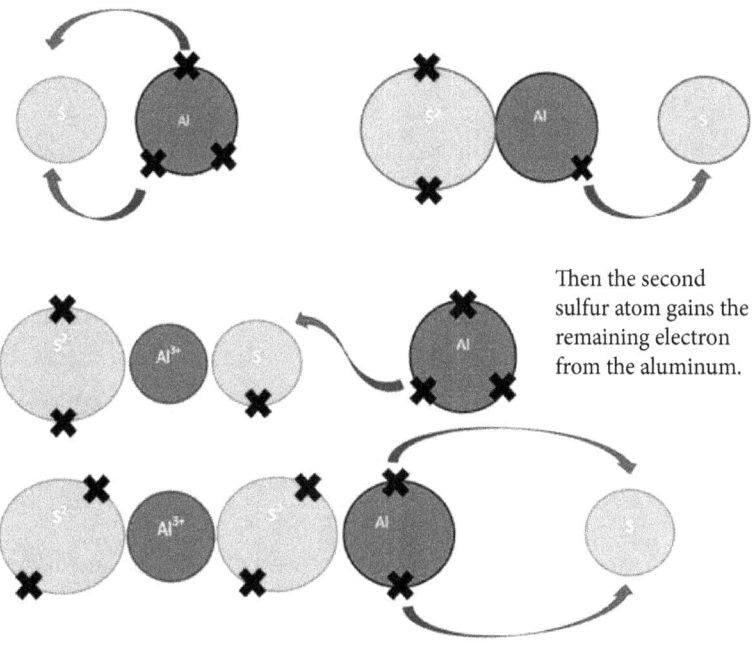

Then the second sulfur atom gains the remaining electron from the aluminum.

Next, a second aluminum atom donates an electron to sulfur atom #2

And finally, a third sulfur atom accepts two electrons from the second aluminum atom. This produces a neutral unit of formula Al_2S_3

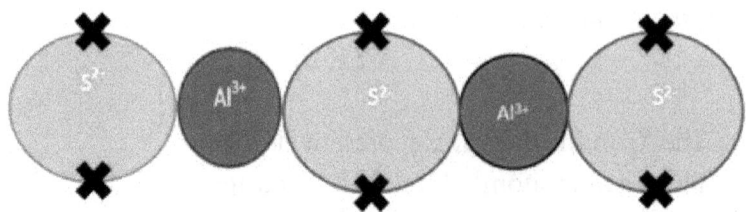

Two atoms of aluminum react with three atoms of sulfur to produce one unit of aluminum sulfide, a recipe for making one Al_2S_3.

So, what if we want to make 2 units of Al_2S_3? We use 4 aluminum atoms and 6 sulfur atoms. Feel free to insert more samples here, if needed, and note the proportional way of thinking. If one substance is tripled, all should be tripled just as when we are making chocolate chip cookies. It is a good idea to write out these proportions so that the students can do so when the problem involves 'messier' numbers. Students find this approach easy as it is intuitive. Most problems which arise are the result of difficulty with previously unlearned material: converting mass to moles or gas conditions to moles, not stoichiometry.

Example: 2 Al atoms/3 Sulfur atoms = 60 Al atoms/ X Sulfur atoms read 2 Aluminum atoms react with 3 Sulfur atoms therefore, 60 Al atoms react with X Sulfur atoms.

What if we want to make 60 units of Al_2S_3? We use 120 aluminum atoms and 180 sulfur atoms.

What if we want to make 60,000 units of Al_2S_3? We use 120,000 aluminum atoms and 180,000 sulfur atoms. Or, written another way, to make 6×10^4 units of Al_2S_3? We use 12×10^4 aluminum atoms and 18×10^4 sulfur atoms.

$2\,Al + 3\,S \rightarrow Al_2S_3$
4   6   2
40   60   20
200   300   100
120000   180000   60000
12×10^4 18×10^4 6×10^4

At this point someone will usually note, "So I'm guessing the next question is, '6×10^{23} units of Al_2S_3,' because that is a mole, which would need 2 moles of aluminum and three moles of sulfur which are the coefficients."

"Right you are! So the coefficients can represent the number of atoms or molecules, or the number of moles of atoms or molecules."

Now provide a couple of sample problems using mole-mole relationships, progress to mass relationships emphasizing that the mass values **must** be converted to moles. (After all, the equation says 2 moles Al with 3 moles S to produce 1 mole of Al_2S_3. This implies 54 g of Al per 96 g S to make 150 g Al_2S_3. Thus, a 2:3:1 mole ratio is far from a 2:3:1 mass ratio). If you have completed the gas laws, you may want to include some problems requiring the Ideal Gas Law. Give them a set of stoichiometry problems (*with answers, of course, so that they can guarantee themselves that they are working correctly*), and you are ready to go to the lab. Note that you have completed the presenting of stoichiometry in one period, or perhaps two if some students are having difficulty with the problems. You should allow some time to assist students having difficulty with the problems during each of the lab days. The second of these experiments requires that the student does stoichiometry for their unique problem. See page 63.

Excess problems are similarly intuitive. Determine which reactant is in excess. Clearly, the reaction stops when one reactant is totally consumed: an analogy with chocolate chip cookies may help here. Then, use the limiting reagent to determine how much of the excess reagent is used, how much remains or how much product was formed, depending on what the question asks.

Many students cite this as the easiest topic of the year. This format also allows a nice transition into equilibrium calculations which employ Initial, Change and Equilibrium quantities. The changes become X, 2X, 3X. -2X, etc. based upon the coefficients and if the substance is being produced or consumed. Those using Dimensional Analysis to solve stoichiometry problems must convert to the proportional change at this point. Of course, we can always make problems a bit more difficult by including percent purity, volume and density for a liquid, etc., but these simply add to the ancillary calculations. The basis for stoichiometry is simple and intuitive. The only areas of common errors are due to the ever-eroding algebra skills and the failure to label quantities, which may lead to an inverted ratio.

Experiments

"Experience is the best teacher."
Julius Caesar

Here are two experiments which I found to be very useful during the stoichiometry unit. We have just finished the reaction unit, so students are familiar with synthesis, decomposition, and replacement reactions, both single and double. The first experiment is an introduction to stoichiometry. It is a variation of an old *Chem Study* Job's Method experiment, probably even older, but I first encountered it there. Students are assigned stations from one to nine for the first nine groups, then 1.5, 2.5, etc. for additional groups as needed so that each lab team has a unique lab station number. They dilute ten times (10 x) station number mL of 1.0 molar copper (II) sulfate solution to a total of 100 mL using distilled water. This contains 'station #/100 moles of copper (II) sulfate. For example, Station 3 would measure 3 x 10 or 30 mL of copper (II) sulfate solution then add 70 ml of water and contains 0.03 moles copper (II) sulfate. This is placed in a double-nested Styrofoam calorimeter (two coffee cups nested in a beaker for stability). They also measure powdered zinc so that the total moles of reactants is 0.10 (so the 0.03 moles group above would use 0.07 moles of zinc or 4.58 g after calculating the required quantity).

Take an initial temperature measurement to 0.1 C using either a thermometer or thermistor suspended within the solution. Add the zinc powder and stir vigorously for five minutes using a stirring rod (*not the thermometer, why?*) while observing and recording the highest temperature during that time. Then filter the mixture, saving both the solid sludge and the filtrate. Place the solid and liquid sequentially (Station 1 to 9) at a site convenient for making observations. This provides students with a visual sequence of results; some residues are black while others are orange. Some filtrates are clear, while others vary from light blue to intense blue. Using a spreadsheet, students share all mass, volume and temperature data with classmates and with students in other classes. This provides a nice collection of points which, when plotted as temperature change versus moles or mass zinc or volume or moles copper (II) sulfate, yields two clear lines which intersect near the center of the range of X values. Plot **all** points from all classes, **not** the average value. If five classes produce the values 4.1, 4.0, 4.1, 4.2 and 15.6, the final point skews the average and needlessly creates confusion. Sharing results allows students to inspect peer data. "Oh, they used the molar mass of the copper (II) sulfate not zinc," or, "They may not have stirred long enough."

As the latter is speculation, ask, "Is there evidence of this?"

Student, "Yes, the solution is still blue and the sludge is black." "They had a huge 50 C temperature change. Did they dilute with the water?" "But water is not a reactant, so why would it matter if the water was used?" Take every opportunity to challenge their logic and help them think more clearly.

The point of intersection, of course, was the major objective of the lab work. What does this point represent? What should the solution look like on each side of the point? Why? What about the sludge? Why? Why do we not see both the copper and zinc in the sludge when zinc is in excess? Can we remove the zinc and see copper? How? (*Remember, they just finished reactions, and a quick check of the Activity Series indicates that zinc should react with hydrochloric acid, but copper will not. Demonstrate with a Station 4 sample.*)

Ultimately, the students create a formal lab report in which they must compare the theoretical results (balanced equation) with the quantitative results from the graph and the qualitative results from the residue and solution observations. The original *Chem Study* procedure only used the quantitative results. I found that the qualitative results made a longer-lasting impression. It also allows the student to visually differ copper metal from the dissolved cupric ion and to differ zinc metal from the zinc ion, another concept with which many students struggle. Students leave this with a much better understanding of limiting and excess reactants, as well as implications of the balanced equation. They also have the foundation for thermochemistry.

The second experiment (which almost immediately follows the above) is one I called 'the two gram lab.' I will quickly sketch the experiment here and give details in an article.[8] Work begins with a list of perhaps twenty insoluble salts. Each lab pair chooses a spokesperson who then claims a salt, first come first served. Each pair is then given an inventory containing all chemicals on site and given the instructions to create 2.0 grams of salt. Not 1.9 grams nor 2.1 grams but 2.0 grams. They must choose their reactants and determine quantities needed, write a procedure, and consult MSDS sheets for safety concerns. The latter is much easier now that we have the internet. You should scan various sources and make recommendations, as some sites have easily read sheets while others are quite complex. Their procedure **must** include a test or set of tests for excess reactants. They must have their procedure checked before proceeding to the lab. Their report must include definition/descriptors of MSDS terms, such as LD_{50}.

Before they begin their calculations, you should remind them of waters of hydration and how these can be compensated for in the molar mass. I also remind them that their reactants must be water soluble if they are completing a replacement reaction. Finally, after reading their procedure, I advise them to wash their product, noting both the technique and the rationale.

This is often cited as the students' favorite lab of the year. It gives them full responsibility for success. When they finish with 1.98 grams of product and just a slight haze of solid when running the excess test, the smile in their eyes is even bigger than that on their lips. They are deservedly proud. It also gives them the opportunity to explore errors. Some finish with 2.10 grams. A short discussion discloses that either 1) they did not wash the precipitate thoroughly or 2) there may be water remaining on the precipitate. Phosphates are notorious for retaining water and will frequently be moist after a weekend of open-air drying. We had two ovens for the moisture problems. Some students chose to slurry and re-filter their sample if it had not been washed. These extra steps may lead to the report being late, but I never reduced the grade for this reason. They were doing what a chemist would do, and that should not be penalized. As an aside, this additional work is done outside of class time, so understand that it will infringe on your prep, lunch and unpaid time. Are we paid by the hour or to do the job right? My average week over the forty plus years averaged ten hours per day at school and perhaps an additional twenty hours per week at home.

When planning the unit, you should include one day for planning as they search for the best type of reaction, complete stoichiometric calculations, etc. Some will want to combine strontium, sulfur and oxygen to produce strontium sulfate. Don't help them directly. That is, don't say, "You need to use a double replacement reaction" but rather, "Do we have strontium metal? How will you control the number of oxygens?" and similar questions. Likewise, some will say, "We want to do a single replacement reaction using strontium and nickel sulfate or potassium sulfate." "How will you separate the nickel from the insoluble strontium sulfate?" After a short while, they will realize that a double replacement is a better pathway. Again, guide them but do not tell them. On the other hand, some may propose a process such as reacting cobalt metal with phosphoric acid. The hydrogen gas will not interfere with product isolation. Don't dismiss this if you have the reactants, as it may work. Isn't that why we call it an experiment? This day or two of small group planning also allows you to identify those students who are struggling with stoichiometry, and you can invite them in for additional assistance.

Other than these two experiments, I spend little time (perhaps two days) on stoichiometry.

Atomic Theory

"Somewhere, something incredible is waiting to be known."
Carl Sagan

For the first half of my career, I taught atomic theory in the traditional manner. Starting with the ancient Greek philosophers ending with Aristotle's Four Elements, then visiting the work of Priestley and others who identified specific elements, determined their relative mass and ultimately created the Periodic Table, thereby overthrowing the Greeks' theory along the way. Finally, we considered the work of Thomson, Rutherford and others culminating in the work of Niels Bohr and the early version of our current theory.

About twenty-five years ago, a very creative cohort, Mr. Dan Rosa, and I turned the educational process around from teacher-centered to student-centered. After introducing the topic and reviewing the work prior to 1860 and presenting some peculiar observations such as the Balmer Series and Rydberg Equation and the clearly non-integer results of some atomic mass values, we split each class into six groups. Each group was responsible for researching and presenting a topic centered around a particular scientist and the resulting change to the atomic theory.

Thus, one group studied J.J. Thomson's work emphasizing his discovery of the electron, including the determination of the mass to charge ratio. The students were told they must gain a clear understanding of and prepare a clear discussion of the latter as the technique would prove critical in the work of Rutherford, Aston and others. We made it clear to this group that they would present first and set the bar for other groups. Furthermore, other groups could come to them during the lab/preparation time to ask questions about Thomson's work and conclusions to gain additional insights regarding their work. Thus, the Aston group could seek assistance pertinent to the mass spectrometer and isotopes, but Thomson could not reciprocate; we work using an understanding of the past, not a crystal ball into the future. Those studying radioactivity might ask why the alpha particle seems to show a mass to charge ratio double that of a proton when, as we know, it has a mass over four times that of a proton. Similarly, other groups could seek insight from previous scientists but not those of the future. Of course, if you refer one group to seek aid from another, you should check back to make certain they received correct information and understand the explanation.

As I circled through the groups during preparation time, I would also plant some questions and suggestions. Why is a computer monitor often called a cathode ray tube (CRT), the name of the device used by Thomson? So Rutherford determined the approximate mass and size of the gold nucleus. Suppose we were to fill a penny with just gold nuclei, what would it weigh? To what is this comparable (a baseball, a person, a car, a locomotive)? How do crime labs use mass spectrometry to complete their work? Students may need coaching on some of these questions but usually rise to the occasion.

I allowed but discouraged the use of PowerPoints during presentations, as I found them restraining creativity. However, students often found other uses for computers during their presentations. Some groups used animations of the Oil Drop Experiment, showing the competing forces and sketching out some of the calculations, for example. Some of the interesting approaches included a group which separated the class members into groups by height (5'1", 5'2", and so forth). They pointed out that some values were not represented, just as some isotopes are not naturally present, and that one could not find the average height by simply averaging the represented heights but must include the occupancy of each; similarly one must include percent abundance in the calculation of average atomic weight. They also represented the separation of ions to the mass/charge principles laid down by the Thomson group. Other interesting approaches included an interview of the scientist by one of the group members while other members demonstrated the phenomenon. (*Interestingly, the young lady who performed the interview became a local newscaster a few years later*). Another group built a cloud chamber and shared videos of tracks left by the radioactive decay.

After a few years of experience, I was able to compress the unit to use one day to organize the groups, describe the apparatus and provide safety information. The following day was spent in the library where I had a number of books placed on reserve. Work with your librarian to create a useful resource center and keep these books available throughout the unit. I also worked with the reference librarian at our local public library to secure copies which could be checked out. The internet is good, but it is difficult to find the depth available in certain other sources. For example, the first third of the text *The Making of the Atomic*

Bomb by Richard Rhodes investigates the development of atomic theory. I highly recommend the book for your background. It is thorough, chronological and easy for the students to understand. After the day of literature search, we devoted four or five days working in the lab and/or preparing for presentations. Six days were allowed for presentations, one day per group. Presenters were expected to take questions as they arose unless their format made that awkward. They would discuss the scientist's experimental design, results and conclusions, emphasizing the changes to atomic theory required by this work.

Students could see a clear progression from the hard spherical billiard ball type model of Dalton to the Plum Pudding Model to a nuclear model with various isotopes and then electron shells, with clear evidence for each change. They also saw the importance of not discarding outlying results (one in ten thousand alpha particles were deflected in Rutherford's Gold Foil Experiment), the importance of remaining open-minded (Becquerel's discovery of radioactivity), the advancement and use of a proven technique versus reinventing the wheel (use of Thomson's mass to charge ratio), and serendipity. Other benefits are the growth in confidence resulting from teaching for an entire period and fielding questions as you do so and the reliance on teammates and group learning, a skill which will pay dividends in college and beyond. Another benefit is that they must remain focused on the presentation of others, as they may be called on to answer relevant questions while members of the audience. For example, if a student asked the Rutherford group how they could isolate the alpha particles without interference from beta particles or gamma rays and the group did not have an answer, the question might be turned over to a member of the Thomson group. This person could review the response of a charged particle in a magnetic or electric field and answer the question.

From my perspective, I saw key advantages to this approach. First, it involved the students very directly with their learning. The retention of their particular study was very strong, as they could still discuss major points a year or two later when they took AP Chemistry. Second, the students could see how knowledge and principles are developed based upon experiments carefully designed to test existing theories. Too often students enter the class thinking that a theory is what a scientist would consider a hypothesis. Finally, many of the current simulations were improved through student input: the project helps foster their creativity.

Of course, all things have two sides (except a mobius strip). The negative issues include 1) time, this approach will require double or triple the amount of time needed for you to simply present the same material, 2) students will remember the scientist who they investigated but not necessarily the others, at least long-term, and 3) you will need to correct and clarify some points for the members of the class. I seldom corrected a group while they were presenting, but instead recorded deficiencies in my notes and addressed those issues later. To do otherwise would challenge their self-esteem and, more importantly, intimidate individuals who had not yet presented. Exceptions were made when an error occurred in fundamental concepts which would lead to confusing results. Ideally, I would pull the group aside, correct their error or at least point out the error and allow them to correct it first within the group and then to the class as a whole, perhaps at the start of the next class period.

Since our approach was rather unorthodox, I would like to take a few pages and describe the work done by students of the various groups. Again, the students were given a key scientist or a set of scientists and asked to discover and understand the contributions to atomic theory. The emphasis in all cases was on the experiment, their results and implications, not on the biography or historical significance of the individual. They start the unit with the Dalton Model, that is, the atom is seen as having the properties: 1) they are hard spherical species, 2) all atoms of a given element are identical in size, mass reactivity, etc., and 3) atoms are fundamental, indivisible and indestructible particles. The work which follows destroys all of those premises except the idea that the shape is spherical, a carry-over from the philosophy of the Greeks.

J.J. THOMSON

The first group focused on J.J. Thomson and his use of the Crookes Tube, which later became known as the Cathode Ray Tube (CRT) and the discovery of the electron. He did so by creating specialized CRTs to explore properties of the ray. For example, he built a pinwheel CRT to determine if the electron had mass. Then he used magnetic and electric fields to determine the mass-to-charge ratio of the particle and to demonstrate that electrons from different metals have the same m/e ratio, i.e. are identical.

We purchased several CRTs including the pinwheel and demonstrated their use to group members, allowing them to recreate many of Thomson's experiments. They did not reproduce his measurements and calculations but understood the principles behind such work and understood why he could only determine a ratio of the two fundamental quantities, not specific values for each. Their class presentation would include these tests, the conclusion by Thomson that this was a previously unknown particle which is much smaller than an atom and the need for a new model of the atom. They discussed their new 'Plum Pudding' Model and the logic behind a model where the tiny electrons or 'corpuscles' were immersed in a positive mass. This model was consistent with knowledge of the day related to Coulombic forces and could easily be defended until Rutherford showed it to be incorrect.

HENRI BECQUEREL

Group number two consisted of several scientists and the term serendipity. The scientists include Wilhelm Roentgen and his chance discovery of X-rays, Henri Becquerel and his unlikely discovery of radioactivity, and Pierre and Marie Curie and their considerable study of radioactivity. To observe some of the phenomena, the group is given a variety of materials. Some groups built and used Cloud Chambers to see the radioactive tracks. Many designed a device to create and then 'bend' a stream of particles and track their path using a Geiger-Mueller Tube. The beam is created using a source fronted by two walls, each containing a pinhole, which effectively blocks particles not aligned with the stream. The level of radioactivity is measured several times and then a magnet is introduced, causing the charged particles to bend. The GM Tube can be used to relocate the particles and perform a count. Removal of the magnet allows a 'background' count to verify that the particles have been relocated. Other groups used paper, cardboard, thin and thick sheets of metal to explore the penetrating power of different particles of radioactivity.

Radioactive sources included an orange Fiesta Ware saucer

found in an antique store. The orange pigment is a uranium oxide and is a strong emitter of alpha particles. Other sources were purchased from scientific supply companies. I used detectors from Vernier, which could be easily interfaced to perform counts. Other groups performed counts which allowed them to calculate and discuss half-lives.

The class presentation includes discovery of radioactivity and types of radioactivity, and the properties of each type as well as transmutation. Since the nucleus was not discovered at this time, they could not describe the alpha particle as a helium nucleus but just as a particle with a m/e approximately twice that of the previously discovered proton. I usually asked the group to conclude at this point and then resume after the presentation by Rutherford when they could explain the radiation as emanating from the nucleus and other work by the Curies. They initially leave with multiple unanswered questions such as 1) how can it be that when we remove an electron from an atom it becomes an ion but when a beta particle (high energy electron) is spontaneously emitted, the atom becomes a new element, 2) how does the element gain the energy released during radioactive decay, 3) why do atoms exhibit radioactivity, and 4) are all atoms radioactive, and will they all eventually decay? This is the only group which routinely returned to the podium to lead a discussion, but the wealth of material and expanse of time over which it was formulated made this approach reasonable. You will see later notes that this could also be done with the last group labeled the Bohr group.

This was probably the least well-developed of the projects, but it also allowed the most opportunity for creativity and exploration. Students who want to be told exactly what to do will be frustrated. Plan accordingly.

ERNEST RUTHERFORD

The third group involved Ernest Rutherford and the Gold Foil Experiment. This is a classic and needs no review.

Here we simulated the work by using a piece of plywood approximately three feet by five feet on which is mounted a tunnel about two inches tall and two feet wide. To it, we attached two bar magnets running parallel to the direction of alpha flow on the underside of the tunnel. A strong rubber band was pulled taut and fixed directly in front of the magnet on the 'source' side of the tunnel, and a ramp was mounted on the source side using a clamp which allowed the ramp to move. The 'alpha particle' was released and could pass through the atom unaffected, it could pass near the nucleus (magnet) and be deflected, or it could approach the nucleus head-on and be shot back at the source. The latter was an unexpected discovery by Rutherford, which caused him to comment, "It was almost as incredible as if you fired a 15-inch shell at a piece of tissue paper and it came back and hit you." Our alpha particle was a ball bearing which was launched down a ramp (your physics teacher should have some of these) and through the tunnel which represents the atom. Marking the board every two centimeters allows the students to systematically scan the atom. By monitoring the point at which the particle exits the board, they can determine if the path was affected as the alpha passed through the atom. Most passes will be unaffected, but as they approach the magnet, they will find the pathway is bent. If you use a stronger magnet, you may find some alphas being captured and will need to be retrieved. The apparatus is leveled, and the tunnel is covered with a towel to hide the rubber bands and magnet supports.

One of the goals is to determine, without looking, what the 'atom' looks like. Their outside work will allow them to relate this to the work of Rutherford, including his determination of the relative size of the nucleus compared to the size of the atom. Encourage them to share the relative sizes. For example, if the nucleus were the size of a basketball, the atom would have a radius of . . . ? Use some common landmarks to emphasize this point, reminding them that the atom is a sphere, not a circle. Why are most alphas unaffected? A second comparison is to ask them what a penny would weigh if it were filled with gold nuclei. Again, find something of a similar mass: a man? An elephant? A locomotive? The Empire State Building? Their presentation should include Rutherford's experimental design, his results, the conclusions drawn as a result and how these conclusions changed the atomic model. Some problems and questions arise as a result of his work, the most important being, "How can the nucleus hold all of this positive charge and not be blown apart?" The answer to this question was impossible before 1905. The change to Atomic Theory was immense. Thomson's Plum Pudding Model was short-lived, as the positive mass has coalesced into a nucleus which is surrounded by a nearly perfect vacuum. Your questions should include, "Why didn't Thomson propose a solar system type model?"

After Rutherford, it is time to return the Becquerel group to the podium to present a better explanation of radioactivity.

Since we had now separated the atom into a nucleus and electrons, I typically proceeded to Francis Aston, the development of the mass spectrometer and the discovery of isotopes, before returning to the electron for the remaining groups. The students seemed to have an easier time if such a path was taken.

FRANCIS ASTON

Our simulation of the mass spectrometer was quite simple. We used a ramp such as that used by the Rutherford group to accelerate the ion represented by a ball bearing. A magnet was mounted such that the ion would pass through the magnetic field, and a system for identifying the landing point was created by the students (some used carbon paper, others marked the impressions on cardboard, etc.). A series of ball bearings was secured, ideally having the same diameter but different mass or iron content (but different mass/diameter spheres of the same composition will work). If you have a skilled tech-ed program, they may help you with this. Students should have a common starting point on the ramp and sequentially roll each ball, marking and measuring the angle of deflection or the distance deflected along an arc. They will observe the same type of interaction observed by Thomson, Rutherford and others. You may wish to hold one ball back and ask them to determine its mass using their apparatus and data.

Isotopes were not discovered by Aston but by JJ Thomson, who observed two well-defined mass values for neon, what we would now recognize as Ne-20 and Ne-22. He apparently passed the information onto an understudy, Aston, who then built a series of increasingly sophisticated instruments and used them to discover most of the known isotopes. Using the instrument, he could determine both the mass and relative abundance of each isotope. Thus, he could accurately determine the atomic weight for the element and address the question of which isotopes are stable and which are radioactive, decay information and other key data. Of course, he also provided the groundwork for the powerful instruments used by analytical chemists today.

Be certain the presentation includes a sample calculation of an atomic weight using an element with at least two prominent isotopes. Your questions for the group might include:

"Why do some common elements like copper and iron have atomic weight values with only two values beyond the decimal point, while others like fluorine have several more?"

"How can scientists use helium-3 to determine if a stone originated in space, i.e. is a meteorite?"

You may also explore questions like, "Why did Dalton think that all atoms of an element are identical, when in fact several isotopes may be present?"

The remaining two groups deal with electrons, so we will leave the nucleus for a while but will return briefly to consider the work of James Chadwick and of Geoffrey Moseley after the student presentations but before going on to electronic structure. Mr. Rosa and I (as well as others) have tried to create meaningful and relevant simulations for these two, as well, but have not been satisfied with results and so have not expanded to include them as a group.

If you have not done so already, you might also note that most of the scientist discussed have studied at the Cavendish Laboratory of Cambridge University including Crookes, Thomson and Rutherford. (*In fact, Thomson studied under Crookes and was the mentor for Rutherford as were Aston, Bohr, Robert Oppenheimer, and William Bragg and a total of seven Nobel Laureates*).

ROBERT MILLKIKEN

In 1910, Millikan designed and completed one of the most beautiful and tedious experiments in the history of science; the

Oil Drop Experiment. This is well-described in *The 10 Most Beautiful Experiments*,[9] another book to add to your collection. This is truly a classic and is still repeated in some introductory physics classes. Our high school students did not have the time, technique nor apparatus to complete such a study. However, we created a simulation which helped them grasp the essential concepts.

We procured a clear plastic tube about six feet long and one and three-fourths inches inside diameter (slightly larger than a ping pong ball). Holes were drilled on opposite sides of the tube every two inches along the tube. These allow some air to escape, thus providing a pressure gradient as air is forced through the tube. A standard hair dryer was attached to a dimmer switch and then mounted at the lower end of the vertical plastic tube. Ten ping pong balls were purchased and a small hole drilled in each so that a copper BB could be added through the hole. One to ten BBs were added to each and the ball was labeled as 10 - # BBs. Thus, the ball containing three BBs was labeled '7' and the one containing 9 BBs was '1.' Students were told that they should treat the number as the number of electrons removed from the oil drop represented by the ball. This way the larger numbered balls representing the more charged oil drops would float higher in the tube given equal air flow just as a higher charged oil drop would experience more repulsion in Millikan's study.

Groups typically chose a constant air flow design, measuring the height at which each drop floated. Others chose to adjust the air flow such that each ball was caused to float at a predetermined height and then the flow rate measured and compared to the charge. Ultimately, they came to understand the importance of each quantity Millikan measured and how he could determine the charge on each oil drop. To help them relate this latter quantity to the desired value, I would give them a simplified scenario. Suppose the charge found on successive oil drops were: 9, 15, 27, 12, 33, 45, 9, 21, 15, 18, etc. Do you notice anything about these results?"

After a moment, someone will observe, "They are all multiples of three."

You, "Right, so we have 3x3, 3x5 ,3x9, 3x4,etc. So what do the 3 and the other number represent?"

"Well, the total is the charge of the oil drop and the changing value must be how many electrons were knocked off so the three must be the charge of each electron."

You, "Right again. Of course, the charge is not three, it is the value in Coulombs which you have found repeatedly. Do you understand why Millikan had to do so many trials?"

"Sure. He had to get enough data to reach an accurate average knowing that each measurement was going to have some error. The actual multiplier wouldn't be as obvious as the '3' in our series."

You, "Excellent. Now what other fundamental quantity did he determine?"

Blank stares between group members, then, "That's all we found about Millikan and atomic theory."

You, "Ah. Go talk to the JJ Thomson group and see what they learned."

We are now ready for the question, "Why was it so important that we determine the actual mass and charge of an electron? Clearly it was important, as many scientists were actively pursuing the numbers and Millikan was awarded the Nobel Prize in Physics for his work.

NIELS BOHR

Some years when my class size became too large (30+), I divided this group into two groups and allowed the first group to present before Thomson, modifying my introduction accordingly. This worked okay but left the first group with a great deal of data and a mathematical formula but no explanation while the second group had to use that data to develop a complex explanation without any real laboratory experience. The data pertains to the bright line spectra of hydrogen observed and measured by early spectroscopist, particularly Johann Balmer. The lines were measured as early as 1885, and three years later the mathematician Johannes Rydberg fit the data to an equation (1/wavelength = 1.097×10^7 m^{-1} ($1/2^2 - 1/n^2$)) where n is an integer greater than 2. These lines can be observed using a simple handheld spectroscope and measured using Vernier's spectrometer.

Later, other spectroscopists working in other regions of the electromagnetic spectrum found similar relationships for hydrogen. For example, in 1906 Theodore Lyman found UV lines fitting the same equation, except the $1/2^2$ term was replaced by $1/1^2$, and in 1908 Friedrich Paschen found a series where the term was $1/3^2$. Later, Brackett and Pfundt found series for 4 and 5. Something very systematic was clearly taking place, but what?

As is often the case, the phenomena could not be explained because at that time we did not possess the fundamental understanding necessary. That understanding first appeared in 1903 with the work of Max Planck. Planck described the corpuscular or quantum nature of light and gave us the equation $E = \hbar\nu$. If you split the group, the second group should review the general results of the spectroscopist and Rydberg and begin with Planck. They must be familiar with this work, as it is essential and the basis, in fact, for their explanation.

Given Planck's equation and the measured frequencies for each series, a set of energy and energy differences can be established. If students are familiar with a spreadsheet, this is a great place to put it to work, as an entire series of calculations can be completed with a few strokes on the keyboard. During presentations, many groups show this beautiful series of values before discussing the work of Bohr, pointing out how the first line in the Balmer Series was the difference in energy of the first two lines of the Lyman series, while the difference between the first two energies of the Balmer Series was equal to the difference of the second and third lines of the Lyman Series and equal to that of the first line of the Paschen Series, and so forth.

When they moved onto Bohr, they typically displayed his equation: $E = -Z^2(k_e e^2)m_e/2\hbar^2 n^2$ (where k_e is the Coulomb constant, m_e and e are the mass and charge of an electron, and n is the shell number) and pointed out that most of the equation was developed using classical mechanics (except the imposition of the integer term). Such terms are uncommon in equations of mechanics, but his calculations agreed very well with the measured values and could not be ignored. But why it worked was still not understood. Why do electrons possess only certain well-defined energies, and why do they not slowly release energy and spiral into the nucleus? As with many advances, the answer to one good question is the development of two more good questions. So in 1913, Bohr (a Danish scientist) used this discovery to cause the evolution of Atomic Theory to leapfrog across the English Channel to Germany, where Irwin Schroedinger, Wolfgang Pauli, Paul Dirac and others took the lead.

If this group is divided into two groups, be certain that the second group begins their presentation with a review of the first group and the questions remaining from the first group before delivering an answer. Similarly, be certain that the first group does not overstep the understanding of the day by sharing our current knowledge of electron shells. They will certainly encounter this knowledge as they complete their preparation, unless they only have access to text written before Bohr. If the study is performed by a single group, I think you could have this group lead off the presentations and then return to complete the explanation (similar to what the Becquerel group did). I never used this approach, but in retrospect, it seems reasonable.

This group must complete many calculations, identify patterns in the results and grasp their significance. As a result, it is the one group which I designed to be non-random. Place students with good math skills and/or good intuition and understanding in the group. These may not be the students with the highest class averages; highly over-achieving students may impede the group. The organization, motivation and follow-through of the latter will prove valuable in other groups. It will also be less obvious to others in the class that you have 'loaded' this group when some of the highest class averages are elsewhere.

Since I am on the topic of group selection, allow me to suggest a few other items. Include both genders in all groups. The teams will work better together than single-gender teams. Place students in groups with an understanding of clashes and comradery: a 'couple' known to have broken-up the previous week are better off in separate groups and even if still together, a dominant member in the couple will frequently suppress the contributions of the weaker partner. Similarly, after forming the groups, review the members: does the group contain several creative risk-takers? Becquerel. Several detail-oriented members? Rutherford. Artistic and creative members? Expect an unusual and theatrical presentation.

A final word: remind the students that a thorough and accurate presentation is expected with the emphasis on the science, not the history and biography. However, they will also be graded on the creativity of the presentation and their personal contribution to the group during both the preparation and the delivery. I provided each student with a rubric describing how the grades would be determined and requesting their input on the project (including evaluation of self and other group members). The results were very similar to the scores I would have given as a result of watching their interactions.

During the presentations, I asked that two groups be ready to present each day. Typically, a presentation with questioning would consume most of the forty-minute period, and the remaining time would be devoted to clarification by earlier groups and updates within groups which had not yet presented. Sometimes a group would be very efficient and clear, such that we could allow a second group to deliver most of their material. On at least two occasions, I had groups which presented and answered questions for two full periods, going into much greater depth than expected but doing excellent work. I also had a reserve of questions which I shared during the preparatory time and some 'stretch' questions to challenge their depth of understanding and application of that knowledge. The latter might include, "Dr. Bohr, astronomers can tell us the composition of interstellar gases which are thousands of light years away. How is this possible?" or "Dr. Aston, you have shown us what the mass spectra of helium would look like. What would happen if the ionizing voltage used were great enough to remove both electrons? Would the spectra change and, if so, how?" Do not penalize the group if they cannot answer these questions, as they go beyond the scope of what was expected of them, but enthusiastically praise correct responses and help them sort through the concepts to reach the correct response if they do not reach it without assistance. If you have fostered critical thinking skills in a safe environment, you will find they are very good at reaching a consensus.

Electron Configuration and the Periodic Table

"He who understands the double slit experiment knows the mind of God."
Richard Feynman

The development of atomic theory unit ends with Bohr's theory of quantum electron shells. The Bohr group ends with the idea that Bohr developed a mathematical model based on classical physics, which yielded an accurate match to observed spectral lines. It did not, however, explain why the integer term was needed. Why do shells exist? It is time for you to take the lead again. The answer comes with the work of French physicist Louis de Broglie in 1924, who proposed that individual particles have a dual nature just as photons do: they are both particle and wave.

Enter Erwin Schroedinger, Wolfgang Pauli, Werner Heisenberg, Paul Dirac and others who developed and applied wave mechanics to explain not only the existence of shells but also subshells, orbitals, electron pairing and other phenomena.

Most, do I dare say all, students have difficulty grasping this concept. I found a series of short videos available on the internet which aided in this mission. These had been produced by the University of Wisconsin and by other sources and were available on YouTube. The videos demonstrate constructive and destructive interference and standing waves using sound and in a wire. They briefly show and discuss harmonics. Other videos in the series demonstrate the same concepts using two- and three-dimensional waves. Seeing these gives the student some insight into what is happening to the electron and why only certain frequencies or energies are allowed. These (or similar videos) are well worth the twenty minutes of class time needed to view and discuss them. Students should realize that these concepts are difficult for nearly everyone. Feynman's quote above seems timely. He was a Nobel Prize winner who worked in this area and was generally acknowledged as a genius.

I developed a thought experiment which, while quite incomplete, gave the students some insight into this dual nature. I used several thought experiments throughout the year and found this one to be helpful. It goes something like this: "Imagine you are a bar magnet. Your head is the north pole, and your feet are the south pole. Now imagine that the ceiling is also a south pole and the floor a north pole. You are being attracted at both ends, correct?" Nods of affirmation. "Now imagine there is a rod, perhaps an axle passing through your middle. We are going to begin spinning that rod. What will happen as it begins to turn?"

Response, "Attractions will decrease, and repulsions will increase."

You, "Good. What will happen as you reach horizontal?"

"Attractions will equal repulsions."

You, "We continue until you are vertical with your head pointed straight down."

"Repulsions increase until your head is pointed straight down if you continue; they decrease again until you reach the original position."

You, "Good. Now what if we had measured the force and plot attractions and repulsions over time, with attractions being a positive force and repulsions being negative?"

"The pen would trace a sine wave."

You, "Good. And if we continue to spin you?"

"The pen continues to trace a sine wave."

You, "Right. Now suppose we have you travel in a straight line as you continue to spin?"

"The pen would trace out a sine wave in the direction I am moving. How fast I move and how fast I am spinning would determine the wavelength and frequency of the wave."

You, "Right again. So, as you travel through space in a straight line, you are creating a wave effect on the surrounding media. This is similar to the electron. This example is much simpler than the dual nature of photons and particles but hopefully gives you some insight as to how both properties can co-exist. Photons and particles are both wave and particle simultaneously. The properties they exhibit are determined by how we observe them: if we look for a wave, we see a wave but when we look for a particle, we see a particle, not a wave. Strange but true. If you are interested in exploring further, read about the 'double-slit' experiment. Now, back to the significance of an electron being a wave . . ."

I usually made some comments about the appearance of spectral lines and what they implied about shells and subshells and how these led to the labeling of the responsible subshells (sharp, principal, diffuse, fundamental). Next, we would dive into the sequence of subshells within a shell, the number of electrons within a subshell and total number of electrons within a shell (as well as the mathematical patterns within each before progressing to orbitals). I found that completing shells before considering orbitals helped students separate the energy factor from location. After introducing orbitals, I would discuss the concept that atoms could minimize energy by distributing electrons spherically about the nucleus. Thus, the combination of orbitals for each subshell should be spherical. This idea was readily accepted, as it seemed reasonable. The shape of the 's' orbital was then trivial, and they could see how the set of three 'p' orbitals described a sphere-like distribution. The 'd's were less obvious but after the first three are seen as symmetrical in three dimensions; we only need examine the last two. The $d_{x^2-y^2}$ places electron density in the x and y direction. A similar orbital in the z and x or z and y planes would result in an unequal (non-spherical) electron distribution in either the X or Y dimension. Thus, the 'donut' of electron density in the XY plane makes sense. I also reminded them that these are not material shapes, but rather regions of space where you are likely to find an electron at a given instant based on wave mechanics. We look at the shapes of the f orbitals, but I did not hold them responsible for knowing the shape of the d or f orbitals. The table of values for shells and subshells can now be expanded to include orbitals. It would look something like the chart below. Typically they can develop the mathematical relationships after the first three shells are recorded.

shell #	1	2	3	4	n
# of subshells in shell	1	2	3	4	n
# of electrons per subshell	2	2,6	2,6,10	2,6,10,14	$(4n-2)$ for $n=1$ to n
Total # of elections in shell	2	8	18	32	$2n^2$
# of orbitals per shell	1	4	9	16	n^2

They are now ready to begin writing electron configurations. If you write the configurations in a pattern similar to the Periodic Table, it will help you tackle a few potential problems. Once you explain what the three terms represent (shell, subshell and occupancy), they will quickly write complete configurations through argon. This is a good opportunity to call on a large number of students in a few minutes. If some are having difficulty, draw some simple concentric circles, placing x's on the circles to represent electrons. Note but don't linger on the fact that you are representing the p subshell as a circle with six vacancies, or draw the p orbitals with two vacancies on each. The latter will depict the complexity of even a rather small atom. The first two electrons go on the first circle, then the 1s subshell is full and we go on to the second shell, etc. Once you finish argon, stop. Ask them to find the pattern (elements in columns have similar electron configurations).

Ask them, "Why did Mendeleyev place elements in the same family?"

Some will answer, "Because they have the same electron configuration!"

Remind them that Mendeleyev proposed the Periodic Table over twenty years before Thomson discovered the electron.

"Oh yeah."

"He arranged them by increasing atomic weight and by similar properties."

You, "That's right. So why do they have similar properties?"

"Because they have the same electron configuration!"

You, "Yes. So now we come to element 19, potassium. What is its configuration?"

Most students will give the reasonable response of $1s^2\ 2s^2\ldots 3p^6\ 3d^1$. However, another student will have picked up your clue and argue that it must be $4s^1$ or it would have different properties than sodium and lithium. The following type of interchange frequently involves various students, although you may have to intercede at times.

"Then where are the 3d elements?"

"I don't know, but potassium is very active just like sodium. Remember the demonstration (or video). Potassium must be a $4s^1$."

"So if potassium is $4s^1$ then is calcium $4s^2$?"

"That would make sense, since it's posted below magnesium."

"Okay, so what about #21, scandium?"

"Maybe that is where the 3ds begin? Is that right?"

You, "Well, what would Mendeleyev imply if there is nothing above scandium?"

"That it is unlike any previous element? Oh, of course, because no previous element has a d subshell."

You, "Well, actually, the shells and subshells are available but unoccupied. Recall how Balmer and the others promoted electrons into higher energy shells and subshells. So, if scandium has a 3d electron, what about titanium?"

"It does, too. So should V, whatever that is, and manganese and iron . . . Let's see, there are 1, 2 . . . 10 of them. Perfect, so those are the d^1 through d^{10} elements, then come the 4p's."

"So where are the f's?"

"They must be down below by Ce. Let's see 1, 2, 3 . . . 14. Yup."

At some point, you should point out that the f block of elements properly belongs in the numerical sequence but has been pulled aside for the convenience of saving space and not covering the entire west wall of your classroom.

Spend a little time bouncing around the table, picking sample elements and noting the electron configuration of the last electron. You may need to remind the students that the d's begin in shell three, so we must count down from scandium as a 3d. Similarly, the f's begin with shell four. They may ask, "Why, if the 3d orbitals are occupied after the 4s orbital, do we not consider them to be 4d electrons?" The theoretician would probably respond that our theory of wave mechanics places them as part of the third shell. This is supported by nodal relationships and energies. I usually explained that, as a result of shielding, the energy of the 3d subshell decreases faster than does the energy of the 4s with increasing nuclear charge so that, for larger atoms, the energy sequence is 3s, 3p, 3d, 4s 4p, 4d, 4f, 5s, etc., justifying the assignment as third shell, not fourth.

If you wish and your students understand the topic to this point, you may want to address inconsistencies at this time. These would include the chromium and copper families. The electron configurations are d^5s^1 and $d^{10}s^1$ not the expected d^4s^2 and d^9s^2 (*How do we know? What is the evidence? You should include both magnetic and spectroscopic evidence)*. Addressing these issues will help them understand why the oxidation states of copper, silver and gold include (1+), which is less common in transition metals than the (2+) charge which results from the loss of both s electrons. There are similar exceptions through the f series and element #46, Pd, is a $4d^{10}\ 5s^0$ not $4d^8s^2$ as expected. I did not address the latter two until AP unless questioned by a student. If you choose to go there, be certain the students understand that these are exceptions to our **man-made** rules for filling of subshells, **not** exceptions to the natural laws to occupy lowest-energy positions first. Similarly, when you discuss the exceptions for the chromium and copper families, emphasize that this is a result of the energy of the nd subshell dropping below the energy of the (n+1)s subshell at this point, NOT a desire by the atom to attain a half-filled subshell, just as the 'stable octet' is driven by energy and not a 'desire' by the atom to fill a shell.[a]

With this approach, the students can write the outermost or complete electron configuration for any atom or monatomic ion without memorization or using the diagonal lines of the Aufbau Method. The position of the Rare Earth elements also seems reasonable.

After quizzing the students' understanding of electron configuration, you are ready for a key concept: shielding. Take time with this to be certain that they understand the principles. These can be simplified to complete 1:1 shielding by inner shell electrons, partial shielding between subshells and lesser shielding between electrons within a subshell.[10,b] Give a few examples such as H, He, Li, Be, B, F and Ne. If they handle these easily, they are ready for three big questions.

The first question is, "So, Kelly, looking at the Periodic

Table, which element has the largest radius, which the smallest, and why?" After a discussion centered on her response and rationale, with opposing views and rationale, the class will decide that Francium is the largest and Helium the smallest, with an upper-right to lower-left increase in size.

Now is the time to show the table of atomic radii and show that their rationale agrees with measurement, at least for s and p subshell atoms. This is also a time when you may wish to have two tables of values available. The first is a standard table which shows numerous inconsistencies. This will allow you to address the question, "How was it measured?" This is a critical question if we are looking at atomic radii or political poll results. Take the time to address an issue critical to good citizenry. Was the data taken in a manner which would naturally skew the results? For example, if a poll is taken where the question is, "Who would you prefer as our leader: our current president or Adolf Hitler?" I can safely say that a majority of Americans would choose our current leader, whoever he or she may be. The standard table of atomic radii is slanted this way, in that the radius is the measured radius of the atom for monatomic atoms or one-half the internuclear distance when two atoms are bonded. Since the internuclear distance for diatomic hydrogen is 0.74 A, the diameter of hydrogen is listed as 0.37 A, not the 0.51 A listed as the diameter of a single hydrogen atom. Now the second table, a table of the diameters of single, monoatomic atoms, can be shared and they find relief in the agreement between their predictions and the table. Inclusion of the first table is optional. I believe that our task is not only to teach chemistry, but also to teach the entire young adult: to prepare them as a skeptical, informed member of society who challenges the pundits of both the right and the left.

Now it is time to address the second big question which is posed to a second student. The question being the same, except that the term 'largest radius' is replaced with 'greatest attraction for its outermost electron.' If this brings puzzled looks, as it sometimes will, rephrase as, "Which element would it be easiest to steal an electron from, and which would be hardest, and why?" Be certain that the answer emphasizes both size and effective nuclear charge, reminding them of Coulomb's Law, which they understand from experience if not from a previous science class. The closer the charged bodies are to each other and the greater the charges, the stronger the force of the attraction or repulsion. Define the quantity as Ionization Energy' and the conditions for measurement.

For both phenomena, we see a diagonal change from upper-right to lower-left positions on the Periodic Table. Be certain that you follow each prediction with a table of values which support your predictions. Be certain that you use atomic radii, not ionic radii. You should also note that, while the patterns exist, there are exceptions, particularly with transition and Rare Earth elements. Also, point out that while the diagonal patterns exist, that does not imply that elements on a diagonal have the same radius or ionization energy. For example, nitrogen, sulfur and bromine are on the same diagonal but do not have the same radius or first ionization energy. However, one can safely predict that sulfur will have a smaller radius than arsenic or selenium and a larger radius than chlorine or oxygen. When examining the table of ionization energies, some students will notice and ask about the inconsistency from nitrogen to oxygen. That is, the IE increases almost linearly from boron to carbon to nitrogen, then drops from nitrogen to oxygen only to rise again in a line nearly parallel to the first as we proceed from oxygen through fluorine to neon so that, if about 200 kJ/mole were added to the latter three, a line would be defined. You may discuss this pairing energy, which they first encountered with Hund's Rule, and please note two other factors. First, if this is a correct analysis, the same trend should be observed when transitioning between the other p^3 and p^4 elements. Is it observed? Why is the effect decreasing in magnitude? The second factor to call to their attention is that the increase from fluorine to neon is nearly identical to the other increases. That is, there is no noticeable support for the idea that 'filled subshells are especially stable.' Yes, the ionization energy of neon is greater than fluorine, but only because the effective nuclear charge is greater. This fact coupled with the fact that neon has no vacancy for an electron results in neon and other Noble Gases being inert (or nearly so).

Now that students understand what ionization energy is, that is the energy needed to ionize a mole of electrons from a mole of gaseous atoms (*as compared to the surface of a liquid or solid or from bonded atoms*), you may continue onto the second, third, and additional ionization energies. Each should increase for a given atom (why?) but more importantly, the magnitude of increase tells us something important. Provide a few examples of successive values without identifying the elements and allow the students to extract the essential factors and suggest the family to which the element belongs. You may wish to start with a rather easy series (such as sodium). The large difference between the first and second ionization energy compared to successive differences implies a weak force of attraction between the nucleus and the outermost electron followed by a large attraction for additional electrons. Thus, the element must be in the IA family. Then perhaps a p^1 or d^1 element such as boron or yttrium or a IIA element can be examined. At some point, another IA such as lithium or potassium should be given and, after identifying its family, ask if the element is above or below the first on the Periodic Table. Why did they predict this placement?

The third big question is similar to the first two. "Which element on the Periodic Table would have the greatest attraction for a free electron and which the least? Why?" This is Electron Affinity. It is easy to predict fluorine as having the greatest electron affinity with the Noble Gases as the least, but the numbers do not support these conclusions. You may hand-wave and partially explain that this is due to the greater repulsion for the additional electron in the small fluorine atom as it becomes an ion and the greater repulsion of electrons by IIA elements than the nearly zero values of Noble Gases. Bottom line, the students' prediction of higher electron affinity in the upper-right corner with decreasing values to the left and down (with the Noble Gases being lowest) is very generally correct, but this process is much more complex than ionization energy. Don't ignore the numbers just because they do not perfectly agree with your predictions. Teach them to test their hypothesis with facts: inconsistencies may indicate that their model is wrong, or it may indicate the model is just incomplete. The facts, however, do not change.

Many textbooks also include periodic trends for electronegativity and ionic radius in this unit. I preferred to introduce these in the bonding unit, although I sometimes used the ionic radii values when addressing electron affinities, as noted above. I think waiting helps the students understand the difference between electron affinity and electronegativity, a distinction with which many students have difficulty.

ᵃ If you and your students appreciate puns, you might note that elements 47 and 79 should be reversed on the table. After they give you a puzzled expression, you can explain that the 4d9ers were in search of gold, not silver. Those with no background in history will still be puzzled, but most will groan.

ᵇ I shared copies of the *JCE* article, "The Genius of Slater's Rules", by James L Reed[10] with all AP students and curious first-year students to provide more rigor and insight. This, of course, was not test material.

Bonding, Shapes, and Polarity

"Chemists don't love, they bond."

Like atomic structure and periodicity, these topics fit together as one and are midway between atomic structure and IMFs and resultant macroscopic properties.

I believe the unit should include electronegativity and that this property belongs in the bonding unit, not the Periodic Trends unit, because unlike atomic size, ionization energy, and electron affinity, it is not the property of an individual atom. Students need to know this and the fact that electronegativity is not a directly measured quantity like ionization energy, but rather is a composite of various quantities as calculated by Pauling. Thus, we have a zero-to-four relative scale as opposed to an absolute scale, such as 2370 kJ per mole for the first ionization energy of helium or a freezing point of 3 K of helium.

To review much of what the students have learned and to focus the students' thoughts for bonding, I enjoyed a game I called 'First or Last?' I would choose a student in a corner of the room and ask, "First or last?" I gave no details until they responded. Then I explained that we would start or end with that student proceeding around the room with each student telling us something about francium. Then we began with students volunteering answers like, "Francium has the largest radius of any element," "It has a molar mass of 223 g per mole," "All isotopes are radioactive," etc. This is something they really get into. When complete, we summarize the key points: a very large atom with low ionization energy and low electron affinity. That is, it is relatively easy to remove an electron and has little attraction for an extra electron. How does this compare to an atom of fluorine? What would you expect to happen if the atoms collide? You may wish to let them 'feel' this by employing pencils as electrons. Have one student hold the electron as francium would (that is, loosely and far from the body) while the second student simulates fluorine (holds the electron tightly and close to the body and has one vacancy in this subshell). As they 'collide' the latter can steal the loosely held electron.

They will correctly predict the formation of an ionic bond, possibly because they learned of it in a previous science class. After requesting their rationale and listing the key points (that is, francium's low ionization energy and fluorine's high electron affinity), it is **necessary** that you examine the numbers. Doing so will show that the ionization energy for francium exceeds the electron affinity of fluorine. Thus, the electron is more stable on francium than on fluorine, and no reaction should occur based on these quantities. But francium reacts explosively with fluorine. What are we missing? Ah, fluorine is diatomic, and francium is a solid metal. Ionization energy is the energy required to separate an electron from a **gaseous atom**, and electron affinity is the energy released by a **gaseous atom** as it gains a free electron, so we must convert each element into individual gaseous atoms. But energy is required to separate diatomic fluorine into atoms and to atomize francium metal. That puts us in a bigger energy hole (more endothermic). Thus, the tendency to react should be less, not more. We are still missing something. Someone (often that quiet, reflective student in the back of the room) will note that the oppositely charged ions will attract each other. Here is your opportunity to introduce crystal formation and crystal lattice energy.

Once the concept of lattice energy is understood by all, it is time to ask, "So if crystal lattice energy is so important, how can a system maximize interactions?" Answer: maximize charge and minimize distance between ions. Thus, calcium becomes positive two, not positive one (both larger charge and smaller ion). Why doesn't potassium do the same? Second ionization energy of potassium exceeds the increased crystal lattice energy (*show the numbers*). Why does oxygen become -2, although the second electron affinity energy is positive? That is, energy is **required** to place a second electron on the oxygen atom (*show both values*). Again, the additional crystal lattice energy of a highly charged anion exceeds the electron affinity requirement. It is energetically favorable to have the second electron on the oxygen anion during crystal formation. Then why doesn't oxygen form a negative three anion? Both size and charge are important. If oxygen were to gain a third extra electron, the third electron would not only experience a much larger repulsive force from the inner portion of the atom (a much larger third electron affinity) than the second. The electron would also need to enter the third electron shell, which would greatly increase its size and reduce the attraction to surrounding cations. This would be energetically unfavorable. This also gives you a good basis for noting that the 'stable octet' is not a driving force, but rather becomes a barrier to further change: calcium +2 would lose a third electron if the additional lattice energy exceeded the third ionization energy. It does not, and so calcium becomes trapped in the +2 oxidation state.

It is time to support the reasoning as mathematically described by Coulomb. Remember, logic is great, but it is not very useful without evidence. This is also a good opportunity to support the Periodic Law. Given the melting point of sodium chloride (801°C), would you expect CsI to be higher or lower in melting point? Why? Give its value. What about KI? KCl? CsCl? Students can predict general trends. Don't expect values to be spot on, but they should understand the trend.

Well, what if we separate the ions not with energy (heat), but by pulling them apart with water? How should the solubility of sodium chloride compare to that of cesium iodide? Why? As always, support your predictions with known values. After they '*Wow!*' at the apparent solubility of cesium iodide compared to sodium chloride, note that cesium iodide is only '**somewhat** more soluble If you are thinking as a chemist.' Let them digest this, and someone may mention moles. Explain that a chemist thinks in terms of particles, not mass. Therefore, we should calculate the number of moles dissolved per liter. The cesium iodide is still more soluble but not by a great amount.

Now is the time for the question, "How should the solubility and melting point of calcium oxide compare to that of sodium chloride? Why?" Again, cite values after making the predictions.

Sometimes a student will comment that one can just use the stable octet to determine ion charges. Before proceeding to covalent bonding, I try again to debunk this myth. To do so, I put together a series of questions related to the myth. These are listed below:

If filled outer shells are extremely stable, then:

1. The first ionization energy of Ne would be _____ the second.
   a. much greater than   b. greater than
   c. equal to   d. less than

2. The first ionization energy of Ca would be _____ the second.
   a. much greater than   b. greater than
   c. equal to   d. less than

3. The first ionization energy of K would be ____thermic.
   a. endo   b. exo

4. The second electron affinity of O would be ____thermic.
   a. endo   b. exo

5. The electron affinity of F is _____ O.
   a. much greater than   b. greater than
   c. equal to   d. less than

6. The first electron affinity of F should be _____ compared to the second electron affinity of O.
   a. much greater than   b. greater than
   c. equal to   d. less than

Before leaving this topic, you should note that most ions, particularly anions, do not have such a simple and easily predicted solubility pattern because of the more complex solvent-solute interactions. You may want to address these coordinate covalent bonds later in the unit, after considering shape, lone pairs of electrons and hydrogen bonding. For most introductory students, these ideas are confusing and can undo the learning that has occurred. Only approach it with students with a solid conceptual base or in AP.

Now it is time to consider the collision of two fluorine atoms (again, a simulation with two pencils is useful). The atoms have a strong attraction for another electron (grab the other person's pencil while holding their own and try to take it away). I emphasized that this 'sharing' of an electron is not sharing in the sense that the United Way or Salvation Army uses the term. Instead, it is a greedy, *'I want to keep my electron and take your electron.'* Since both atoms do so, they 'share' the pair of electrons. In this case, the sharing is equal. Notice that you have prepared them for the concepts of electronegativity and polarity.

How many bonds will an atom make? Answer: as many as it can. Why? Be certain that you address energy here.

Thus, oxygen usually makes two covalent bonds, as it uses two half-filled orbitals to overlap with two other half-filled orbitals (be they on a single atom as when an oxygen atom forms a double bond with another oxygen atom or with two other atoms as in water or an alcohol). Similarly, nitrogen usually forms three bonds. However, this is not necessary, as either can fill an orbital with an electron (as in the hydroxide ion) or use a lone pair and an empty orbital to form a bond (as in the hydronium or ammonium ions or when boron bonds with nitrogen during the reaction of boron trifluoride with ammonia). **Stress** the fact that energy is **released** each time a bond **forms**, and energy is **absorbed** when a bond is **broken**. The difference in these two processes determines if the reaction is endo or exothermic. It also is important in determining the stability of the substance. If a bond is easily broken, change is inevitable. Including species such as hydronium and ammonium also sets the table for discussing expanded octets and coordination compounds in later courses or with more advanced and curious students.

I have only a few other comments which may be helpful. The first has to do with expanding the idea of bond strength and stability. I include this because it can then be used to justify instructions in the process of drawing Lewis Dot structures. We begin by looking at the trend in bond energies, such as C-C vs. N-N vs. O-O and in a series such as C-F vs. C-Cl vs. C-Br vs. C-I (show the numbers, of course). We then demonstrate the effect of the weak O-O single bond using peroxyacetone and the weak N-I bonds of nitrogen triiodide. After doing so, be certain you emphasize the important role of new bond formation: converting nitrogen triiodide into a nitrogen atom and three iodine atoms would be endothermic and unfavorable, but formation of diatomic iodine and especially the strong nitrogen-nitrogen triple bond of diatomic nitrogen makes the process very exothermic (also note that much of the energy is released as sound). *IF YOU CHOOSE TO USE THESE DEMONSTRATIONS, DO NOT SCALE THEM UP. USE SAMPLES NO LARGER THAN THOSE DESCRIBED BY SHAKASHARI.[11] LARGER SAMPLES CAN BE POWERFULLY EXPLOSIVE!*

Early in my career, I found that many students had difficulty drawing Lewis dot structures when trying to follow the sequence given in textbooks. If given a formula such as NH_3O, they might begin with nitrogen and then add the three hydrogen atoms, leaving no space for the oxygen atom. After observing their struggles for a few years, I developed a sequence to help them. It took the steps:

1) Begin with the atom which makes the most bonds, that is, has the most half-filled orbitals.
2) Proceed to those which make fewer bonds.
3) Before placing single bond species, count the number of vacancies remaining. Remember to add or remove electrons if you have an ion.
4) If you have more vacancies than single bond species, begin forming multiple bonds, especially using two second period elements. Some students will also note that cyclic compounds may be possible.
5) When the number of vacancies equals the number of single bond atoms, place the atoms.
6) Check for isomers, remembering that any atom which makes two or more bonds can be inserted between two other atoms of the structure. Why?
7) Avoid O-O bonds and N-N bonds if possible.

After they have developed their skills, let them build models to see the 3D structures. This is also a good time to show the difference between a true isomer and a 'paper' isomer. Build a couple of models and show how the latter are the same structure simply twisted around in space, while isomers require a change in the bonding pattern (example, ethyl amine versus dimethyl amine).

You can again emphasize the significance of the weak O-O and N-N bonds by discussing the instability of hydrogen peroxide and why it is kept in a dark bottle in a cool location. Again, you can examine the bond strengths for products and reactants. Drawing nitrogen dioxide and dinitrogen tetroxide shows a weak nitrogen-nitrogen bond vs. two unfilled orbitals. Demonstrate by trapping nitrogen dioxide in a syringe (***CAUTION: NO_2 is corrosive and toxic! Produce and destroy in a fume hood***). By applying pressure, you can force the colored NO_2 to experience more collisions, resulting in the formation of more colorless N_2O_4. Students can visually see the color changes as you compress and expand the gases. Placing the syringe into ice water or hot water will cause similar changes. These are also common demonstrations during the thermodynamics or equilibrium units of AP or General College Chemistry.

At this point, the student should be able to write Lewis Structures and address bond polarities. If you choose to discuss 'sp' hybrids (*I did, but some teachers do not*), this a good place to do so. (*I found the overhead images provided with Zumbdahl's text[12] to be particularly useful*). Doing so helps the students understand why multiple bonds count as one region when using VSEPR Theory. The latter will allow the student to predict the geometry about any atom in a molecule. Have the students draw a few more structures involving several pi bonds, predict the geometries and then build and show the structures. Don't be surprised when that 'average' student from the back of the room, the one who grew up with Legos®, takes control of the discussion, turning heads with his or her insights and 3D expertise when talking about molecular shapes and polarity. When discussing shapes, you will find it much easier if you use the terms 'with respect to orbitals' and 'with respect to nuclei.'

You can now address molecular polarity using the previously built molecular models. This becomes the obvious lead into intermolecular forces, solution formation and other macroscopic properties.

IMFs and Solution Formation

"Few scientists acquainted with the chemistry of biological systems at the molecular level can avoid being inspired."
Donald Cram

One of my favorite lessons of the year begins with, "Everyone, imagine that you are a molecule on the surface of a liquid." Pause. "Bill, you are about to evaporate." Pause. (*Sometimes, a creative Bill will evaporate. You may follow up with, "Unfortunately, after leaving the surface you undergo an elastic collision with a diatomic nitrogen molecule from the atmosphere and return to the surface of the droplet, where you again become part of the liquid."*) "What makes you special?"

After a moment of thought, Bill will give his response with mandatory rationale. Then, "Mary, do you agree or disagree? What would you change?" And so, the discussion develops, and soon they have fleshed out the key points: Bill must be moving in the right direction and with sufficient energy to overcome the attraction of surrounding molecules. You may remind them that we saw the distribution of velocities during the gas unit. The direction of motion is basically random, so we must focus on the issue of attractions. The table is set. Go at it.

The topic of IMFs was discussed quite thoroughly for ionics during the previous unit, so I only did a cursory review. Dipole-dipole forces are quite similar to the ionic forces but weaker due to the lesser charge on the molecules compared to ions. London or dispersion interaction forces, however, are more difficult for them to grasp and master. Diagrams and discussion of the effects of net nuclear charge and size on the polarizability of a species will help them grasp the concept and make crude predictions of the resulting properties. This is a good place to discuss the condensation and solidification of gases and to review the Ideal Gas Law with its correction factors.

Perhaps the most difficult force for them to understand is the hydrogen bond. I think the semantics are a chief culprit. I had much better success when I referred to it as a hydrogen dipole, which has the technical name of a hydrogen bond, but which is really an IMF containing hydrogen. As with other intermolecular forces, its strength is defined by Coulomb's Law: the more highly charged (in this case, partial charge) and the closer the two species can approach each other, the stronger the force of attraction or repulsion. Since the bond involving hydrogen is less polar for H-N than for H-O (which in turn is less polar than H-F), the hydrogen bonding follows the same sequence. They usually follow this easily but then are stumped by the question, "So why are the boiling points for the simple compounds $NH_3 < HF < H_2O$?"

This gives you an opportunity to return to first principles and exam the tetrahedral shape of each. Someone will usually spot the fact that water can form hydrogen bonds off each of its four orbitals while the other two can, on average, only form hydrogen bonds with two of the orbitals due to a shortage of either lone pairs or polarized hydrogen atoms. A table or chart of the boiling points of the hydrides of the carbon, nitrogen, oxygen and fluorine families will emphasize the significance of the hydrogen bond. You can further emphasize this by comparing boiling points of similar molecular weight compounds, such as methanol and ethane, ethyl alcohol and propane, or ethylene glycol and butane and pentane or the dramatic increase from propane to propanol to propylene glycol to glycerol. Overcoming all three hydrogen bonds in the latter is very difficult, resulting in a low vapor pressure and a high boiling point. As always, support your arguments and predictions with facts and actual values. You may also wish to strengthen earlier concepts by asking what will happen to the boiling point if the liquid is warmed in a low pressure (vacuum) system. Why? This will give you an opportunity to explore the process of boiling and phase diagrams. Your AP Biology teacher will also appreciate students who enter the class with at least a rudimentary understanding of hydrogen bonding and the ability to recognize it.

You now have the two key concepts in place for acid and base chemistry. First, of course, is the strong hydrogen bonding which exists between molecules containing N-H, O-H or F-H bonds and a second species containing one of these bonds, be they different species such as ammonia and water or two of the same species such as two ammonia molecules. The hydrogen bonding can also occur within a single molecule as in amino acids. The second concept is that hydrogen bonding can result in the ionization of a bond to produce hydronium or hydroxide ions, which are the basis for aqueous acids and bases, and this is a primary reason for the differing strengths of acids and bases. Before turning our attention to these acids and bases, we must take a brief look at solution formation, solution concentrations and colligative processes.

SOLUTION FORMATION

I have indirectly addressed the topic of solution formation during the ionic unit and above, but I must go into a little more detail. Textbooks often use the phrase, 'Likes dissolve likes,' without addressing why this is true. The answer is intermolecular force. We saw how the polar water molecules could surround and attract cations and anions, pulling them into solution where additional layers of water molecules could attract them and keep them in solution. Competing with this process are the solute-solute and solvent-solvent attractions, that is, the Coulombic attractions between ions and between polar solvent molecules must compete with the forces attracting solute to solute and solvent to solvent. At some point, the two processes reach a balance which we call equilibrium (in AP you can address the enthalpy/entropy aspects of this equilibrium). Since nonpolar solvents exhibit, by definition, very little polarity, they will be very ineffective at overcoming the ionic bonds and will show very low solubility for ionic compounds.

If our solute is a polar molecule and we use a polar solvent, we would expect significant attractions between the two and some (perhaps extremely high) solubility. This is particularly true if both or neither are capable of hydrogen bonding. Thus, we find substances like methanol, ethanol, ethylene glycol, glycerine and similar substances to be miscible with water. Other substances (such as butanol or octanol) are partially soluble in water. The reason for the decreasing solubility is not that the water to hydroxyl group attractions are becoming weaker, but it is due to the fact that the nonpolar carbon-hydrogen chain is interfering with the hydrogen bonding between water molecules. Thus, solubility will tend to increase as we change solvent from water to methanol to propanol and beyond. The same is true if we use a nonaqueous, polar solvent such as liquid hydrogen fluoride or methanol. If a polar solvent is used which does not hydrogen bond, such as chloroform, it will serve best as a solvent for other polar, non-hydrogen bonding solutes, because the solute-solvent interactions will be very similar to the solvent-solvent and solute-solute interactions.

At this point, many students will claim that nonpolar solutes like gasoline or octane are not soluble because the very polar water molecules cannot attract the octane molecules with enough force to separate them from each other. Ask them to reason this through. We know the octane molecule is nonpolar and so has only weak induced dipole forces of attraction for the surrounding octane molecules. Thus, it is quite volatile despite a rather high molar mass and long threadlike structure which can twist around neighboring molecules. That is to say, very little force is needed to overcome these attractive forces. If we replace some of these octane molecules with the much more polar water molecules, what effect should we expect for the attractions? Well the water should create a greater polarizing effect on the electron cloud of the octane than does a second octane molecule. This, in turn, should result in a stronger attraction between the water and octane than between two octane molecules, which suggests that octane should be soluble in water. But it is not.

Why not? What are we missing? It is time to go back to our earlier statement about interactions. We have three types of interactions but have only considered two: solute-solute and solute-solvent. What about solvent-solvent? The water molecules are hydrogen bonding to each other. When you introduce an octane molecule, the hydrogen bonds of the water molecules must be overcome. This requires energy input, more energy than is released due to the formation of the induced dipoles of water and octane. Thus, octane and other nonpolar substances are not soluble in water or other highly polar solvents. This is true, not because water cannot pull the nonpolar solute molecules apart, but because the water molecules cannot be pushed apart by the solute molecules. This is a seemingly subtle difference but is important in the development of consistent thought.

Here is another opportunity for 'experiments at home.' Ask the students to go home tonight and mix a little vegetable oil with water and shake the mixture vigorously, then observe the result. They will see the oil collecting on the walls, sides and bottom of the container and then slowly release and rise to the surface. The droplets were being pushed aside by the water.

Before leaving the topic of nonpolar solutes, you should take a minute to point out that shape can also be important. Thus, a long, 'stringy' molecule like n-octane will be a better solvent for some substances than is the nearly-spherical dimethyl propane or the flat benzene molecule. The more similar the shapes, the more soluble the substances are in each other, all else being equal. Thus, the line, 'Likes dissolve likes.'

You may have noticed that thus far we have only considered dissociation. We have looked at the separation of ions from a crystalline lattice and covalent molecules from each other. We have not considered ionization, that is, the formation of ions during solution formation (such as we observe when hydrogen chloride gas dissolves to form hydrochloric acid). We shall do so very soon.

SOLUTION CONCENTRATIONS

The concept of solution concentration is intuitive to most students. The idea that there may be multiple ways of expressing concentration is not always so. I usually began with a discussion of what would be a reasonable method of expressing concentration. Most quietly accept that we are primarily focused on the solute. We express the amount of solute per solvent or per total solution. If questioned why we do so, you can ask them, "If you are considering the drinking water in your home, are you interested or concerned with the amount of water per glass or the amount of lead, Atrazine or other contaminant per glass?"

Then take student suggestions for possible concentration units. Typical suggestions include g solute/g solvent, g solute/g solution, g solute/L solvent, g solute/L solution, moles solute/L solvent or solution, moles solute/Kg solvent or solution, moles of solute/(moles of solvent + moles solute) (clearly explain why this is not moles of solution). Be accepting of any viable expression. Stay positive with their emotions. They may propose a viable expression which we do not use simply because it is not useful in any of our more sophisticated theories. They do not know this yet.

Now you can begin to pull out the expressions which chemists have found to be useful and point out some of the ways in which these are frequently used. For example, biologists and chemists frequently use the second expression, slightly modified, to give per cent strength. Analytical and environmental chemists use a slightly different version to express ppm's or ppb's. The molality expression is useful when studying colligative properties and, of course, molarity is the workhorse of the group used repeatedly in numerous areas (such as stoichiometry, pH, kinetics, equilibrium, electrochemistry, and the list goes on). This also gives you an opportunity to discuss the analytical chemist and how she might work as a toxicologist, environmental chemist or engineer, crime lab investigator, or quality control chemist in an industrial setting. Feel free to circle or asterisk the expressions, adding the label and any modifications so that they have a set of expressions including percent strength, molality, molarity, mole fraction, and density.

Give them a couple of sample problems and ask them to complete the calculations so that they can comfortably differ the terms solute, solvent and solution. Example problems might include 1) 40.0 g of barium chloride are dissolved into 1000. ml of water producing 1020. ml of solution. Calculate the density, molarity, molality and percent strength of the solution. 2) 500.0 ml of a 1.00 M KBr solution with a density of 1.05 g/ml are dehydrated and 490. ml of water are collected. Determine the mass of solute, mass of solution, molality, mole fraction, percent strength and density of the solution.

Point out that the chemist is usually focused on the particle relationships, and thus focuses on molarity, mole fraction, and molality. Since all 1.0 M solutions contain 1.0 mole of solute per liter of solution, they can compare properties of a 1.0 M sodium chloride in water solution to a 1.0 M octane in hexane solution, or in a similar way, 1.0 m solutions which use the same solvent. For example, if one compares a 1.0 m sodium chloride water solution to a 1.0 m sucrose water solution, we know there are vast differences. The first is an electrolytic solution, the second a non-electrolytic solution, the first tastes salty, the second sweet, and so forth. However, by definition, each contains one mole of solute per 1.0 Kg of water. Since water has a molar mass of 18.0 g/mole, this implies we are using 55.5 moles of solvent. Thus, the solutions each contain one molecule (or formula unit) of solute per 55.5 molecules of solvent: a useful fact which allows us to look for patterns of properties. If we examined 1.0 m solutions when hexane is the solvent, we would find 1.0 mole solute per 11.9 moles of solvent. Consequently, patterns we find for molality also exist, with numerical adjustment, for mole fraction. This is not true for molarity, as the solute occupies some of the volume, reducing the quantity of solvent present. So, comparing 1.0 M saltwater solution to a 1.0 M sucrose water solution, we may find the larger sucrose molecules occupy a greater percentage of the total volume, such that the sugar water solution contains less water than the saltwater solution. You need not share all of this with your students, but **YOU** should understand it. I have heard non-chemistry majors make statements such as, "Yes, molarity and molality are essentially the same." While this may be true for very dilute water solutions, since a dilute solution is primarily water and one liter of water has a mass of one kilogram, it is not true for concentrated solutions and is certainly not true for nonaqueous solutions.

A very useful lab exercise at this point is to bring out a few solutions from the stockroom and note that the identity of the solution is given but not the concentration. Their assignment is to determine the concentration by any method **except** solvent removal/evaporation. The creativity is fantastic. Some, of course, approach using stoichiometry and gravimetric techniques. If the solute is copper (II) sulfate, they might form barium sulfate or copper(II) carbonate. Or they might use powdered zinc followed by hydrochloric acid to remove excess zinc and then filter off the copper metal. Others made a series of solutions, striving to match the intensity of colors. Others used a colorimeter or plated the copper using DC electricity. Still, others employed an analytical balance and volumetric flasks to create a density gradient against which the solution could be compared. Some battled the challenge of determining an accurate freezing point for the solution. When finished, we compared results and discussed which results were probably the most reliable and why. This is a simple lab with wonderful opportunities for growth by the students and you. These simple forays often gave me insights into other experiments which could be developed.

COLLIGATIVE PROPERTIES

I introduce this topic, as with most topics, with a series of questions. "Have you ever left soda or beer in the garage in the winter? (*I am from Wisconsin.*) What happens?"

Excited stories. "Yeah, we got up in the morning and found exploded bottles all over. Frozen stuff."

"We once had Diet Coke cans explode, and regular Coke cans right next to it didn't."

"Yeah, my dad had the same happen with light beer but not the regular beer."

You've set the stage, "Why did the Diet Coke freeze but not the regular?"

"Maybe sugar versus sweetener?"

"Maybe the amount of sugar and other stuff in solution?"

"Maybe the color?"

"No, I've seen the same thing with 7UP."

Let them continue, as long as most of the comments are moving you forward. Finally, you interject, "Okay, how could we test your theories?"

"Make a series of solutions by diluting a Coke, and then see what temperature each freezes at."

You, "Does it have to be Coke?"

"No. Okay, make a series of solutions and see what temperature each freezes at."

You, "How would we do that?"

"Just weigh out different amounts of sugar and dissolve each in 100 ml of water. Set outside on a cold night and check the thermometer when each begins to freeze. I suppose you could use a thermistor so that you don't get cold."

You, "Great. That is exactly what some scientist did, and here are their results." Present a table such as that in Appendix 3.

After presenting this data, point out that all solutions have been prepared using 1.00 Kg of water. That is, the molality of the solution is numerically the same as the number of moles of solute. Then step back and let them go as far as possible on their own. I once had a group work their way through all three sets of data (covalents, ionics and acids both weak and strong), picking off first the general equation for methanol in water (FP = -1.86 m), then recognizing that it worked for all of the covalents and interpreting this to mean that the size, shape, polarity and other characteristics of the solute have no effect and that the freezing point only depends on the molality or solute-to-solvent particle ratio. They then picked off the integer multiplier (van't Hoff factor) for ionic compounds, correctly relating it to the number of individual ions placed in solution as each unit dissolves. Finally, they collaborated to see how the same principle applied to acids if we assumed that some acids ionized completely while other acids only partially ionized. I sat to the side in awe as they worked through the sequence and pulled out both patterns and rationale. Finally, when all were in agreement, they turned to me and asked, "Is that right?"

My response after a moment of thought was, "I don't know. How could we test your thoughts on the acids?"

After a few troubled glances, one suggested, "Well, if one makes lots of ions and the other just a few, the first should be a good electrical conductor, and the other not."

We immediately grabbed a conductivity devise, a bottle of dilute hydrochloric acid and equally dilute acetic acid. A few moments later, the room was abuzz with cheers and high-fives. This is the reason we teach. Please keep in mind that this occurred once in perhaps thirty years, with each year including multiple first-year classes. Don't expect such results immediately, but gently guide students through the analysis, digesting small quantities of data at a time, looking for patterns within the data, sharing ideas and insights in a safe and collaborative manner, and proposing reasons which are consistent with their current knowledge. It is not surprising that several members of that class are now doctors, engineers or research scientists. Many groups discovered the first equation, and a few interpreted it. A few discovered the trend for ionic compounds. Only one went further without assistance.

Many scientists who possess more knowledge than I do would criticize this approach, arguing that entropy, activity coefficients and other factors should be involved in the discussion. I argue that there will be time for that after the students have developed better math skills and the fundamentals of thermodynamics. My role is to pique their interest, give those for whom General Chemistry is a terminal science course a better understanding of how science works, and to assure those who are going to pursue science that this is certainly not the complete picture. Should a physics student begin his or her study of motion with an examination of relativity and quantum mechanics? Are these not the theories which best describe our current knowledge of large and small particle motion?

When you finish this exercise, inform them that while the general principle is correct, some of the values were false, and that the exercise was created to help them discover these general principles. If they ask about specific 'fake data' you might point out that most ionics are not one hundred percent dissociated at any given instant. For example, the complete ionization of $Fe(OH)_3$ accounts for only a trace of its actual solubility.[13] Some covalent compounds listed may not be soluble to the extent cited or may react with the water, rather than simply dissolving as the table would imply. I chose representative substances based on their elemental composition, number of ions, etc. to illustrate the pertinent factors. Similarly, the recorded freezing points of the weak acids do not reflect their actual percent ionization but are, again, only for concept development. Actual freezing points for some weak acids would not vary by more than a few hundredths of a degree, and the van't Hoff factor would not be discernible.

Solution Stoichiometry

"An experiment is a question which science poses to Nature, and a measurement is the recording of Nature's answer."

Max Planck

One of the advantages of teaching one subject full-time for a number of years is that you can work different units together in a nice, consistent pattern to help students carry their understanding from one topic to another. For example, I pointed out earlier how understanding that a mole represents a certain number of particles just as a dozen represents twelve objects, helps students recognize that the coefficients of a balanced equation can represent the ratio of atoms, molecules or other reacting species but can also represent the ratio of moles of reactants or products. This understanding can now be used for solution stoichiometry. Here is how to do so. I will again use a pseudo-discussion format.

You, "John, suppose 365 g of hydrogen chloride were dissolved into enough water to create 10.0 L of solution. What would be the molarity of this solution?"

After some calculations by John and the rest of the class (they know you will ask them if they agree with John), "1.00 M."

"Agreed? And what if we dissolve 800. g of NaOH into enough water to prepare 20.0 L of solution?"

Mary, "1.00 M."

"Right again. Now, what would happen if we mixed 1.00 L of the hydrochloric acid solution with 1.00 L of the sodium hydroxide solution? Yes, Bill."

"We would have 2.00 L of solution."

"True, but what else would happen?"

"I suppose they would react because they are making water as a product. Oh, and heat would be produced."

"Right on. Here, show us the balanced equation."

Bill does so and comments, "Oh, that's just water and table salt, so it is no longer an acid. Let's see. Okay, it is already balanced."

"All agreed?" Pause. "Good, so Jessica, what does the reaction tell us?"

"It says that one mole of hydrochloric acid reacts with one mole of sodium hydroxide to form one mole of water and one mole of sodium chloride."

"Perfect, but we know that the two reactants were in solution. How can we account for that factor?"

Jessica again, "Well, they are each 1.00 molar, which means there is one mole of solute in one liter of solution. So, if we used one liter of each solution, we would have used one mole of each reactant, so we would have just the right amount of each reactant. We used up all of the acid and the sodium hydroxide."

"Exactly, now what if we had used 1.00 L of 2.00 M HCl instead of the 1.00 M HCl, what volume of NaOH would be needed?"

Experience says that, after a brief pause, about 80 percent of the students will answer 2.0 L while the others respond .50

L. The latter comes from the thought that 1.0 mole of solute in 0.50 L gives you a 2.0 M solution. That is not the situation, however. It is time to walk them through the logic of the problem. 2.0 M implies that you have 2.0 moles of solute in each liter. Since you have one liter of solution, you clearly have 2.0 moles of HCl. The balanced equation tells us that we need 2.0 moles of NaOH to completely consume the 2.0 moles of HCl. The 1.00 M NaOH tells us that we have one mole of NaOH per liter, so to obtain 2.0 moles of NaOH we will need 2.00 L of solution. To assist them further, I set up a table such as this:

$$HCl + NaOH \rightarrow HOH + NaCl$$
M L M L
1 1 1 1
2 1 1 2

Then expand changing the unknown, perhaps like this:

2 2 1 ? 4
3 ? 2 2 3
? 4 3 6 2

Finally, .117 .0251 .164 ?

It is time to develop an equation, which they quickly see is $M_a V_a = M_b V_b$ which they can use to solve the latter problem.

"Good, now what about the reaction $H_2SO_4 + NaOH$?"

Have someone come to the board and balance the reaction, then repeat the numerical sequence. If there is disagreement or an incorrect response, repeat the logic from the definitions and stoichiometry. A table such as that below will emerge.

H_2SO_4 + 2 NaOH → Na_2SO_4 + 2 HOH
M L M L
1 1 1 2
2 1 1 4
2 2 1 8
4 3 6 4
1 3 2 3

They will quickly develop the equation $M_a V_a$ / coefficient$_a$ = $M_b V_b$/ coefficient$_b$ which can be used for any solution stoichiometry problem. Be certain that you point out that M x L$_{solution}$ gives you moles of solute so that you are really just using a variation of the equation developed during the stoichiometry unit. If a problem gives a quantity in grams, we can easily determine the moles of that substance and use that information to determine the volume or molarity of solution with which it would react. For example, if 29.0 ml of a phosphoric acid solution completely consumes 0.560 g of potassium hydroxide, what is the M of the acid?

You have probably noticed that the reactions used here are neutralization reactions. That is because I would be moving on to the acid base unit next. In fact, this topic (solution stoichiometry) often coincided with the arrival of spring break. Consequently, some years I would use it to conclude the solution unit and other years as an introduction to acids and bases. Practice problems should include non-acid base problems, some of which may recall the 2.00 g Lab or other earlier work. For example, "During the 2.00 g lab, some groups used a solution containing 40.0 g of $BaCl_2$ in 1.00 L of solution to form 2.00 g of barium carbonate. What was the M of the solution, and what volume of this solution was needed to give the desired yield?"

You may want to revisit this general equation if you include stoichiometry in the redox unit.

Acids and Bases

But does it pass the litmus test?

This is a topic which the students approach with excitement but also caution. They have heard about acids and bases for a long time. They have used acids and bases in several of their earlier courses and can recite many properties of each: acids are corrosive, have a pH below 7, neutralize a base, turn litmus red and so forth. Having just completed solution stoichiometry, acid base neutralization seems like a good place to begin this unit. I believe that all students should be asked to use a burette to carry out a neutralization reaction. The fine-motor skills, the patience, the observation of indicator sensitivity as the mixture turns bright pink and then clears and the satisfaction of success cannot be duplicated with a computer simulation. Each team should complete several titrations to determine the concentration of an unknown solution and to produce precise results.

The topic of ionization of acids, that is, the production of the hydronium ion and an anion, was introduced during the colligative properties unit. At that time, we saw that some acids such as HCl, HI, HNO_3, H_2SO_4 and $HClO_4$ ionize 100 percent in water and are considered to be strong acids, while others such as HF, $HC_2H_3O_2$, H_2SO_3, HNO_2, and H_3PO_4 are considered weak because they are only partially ionized. If you prepare a table of these and their percent ionization in a 1 M solution, the students will quickly spot several factors which affect the degree of ionization (see Appendix 4). The key determinants for oxyacids are 1) the number of oxygens on the central atom: the more oxygens, the stronger the acid, 2) the more positive the species, the stronger the acid: each successive proton is more difficult to remove, 3) the more electronegative the central atom, the stronger the acid.

Great! But why? What is the common thread to this trend? This is a nice topic to allow students to 'feel' the answer. Have them draw an electron dot structure for a series such as the $HClO_x$ group. You may have to remind them that oxygen is very good at forming coordinate covalent bonds by simply forming an empty orbital and overlapping that empty orbital with a filled orbital of the central atom, in this case chlorine. Oxygen is particularly good at this, as it has a high electronegativity, second only to fluorine. Thus, not only is oxygen using chlorine's electron without donating an electron, but it also has a greater affinity for the bonding electrons than does chlorine. The net effect is that electron density is pulled away from the chlorine and towards the oxygen, causing the chlorine to become effectively more positive. This, in turn, causes the chlorine to attract the electrons of the Cl-OH bond more strongly, causing that oxygen to become more positively charged which, in turn, causes the hydrogen to become more positively charged, thereby forming a stronger attraction to the solvent, water molecules, resulting in greater ionization. Additional oxygens intensify this effect.

To demonstrate, ask for a volunteer. Give this individual three pencils, two in one hand and one in the other. Explain to the class that the pair of pencils represents one of the lone pairs of the chlorine in their drawing. The single electron is bonded to an oxygen atom (second volunteer with one pencil in each hand representing the two half-filled orbitals of an oxygen atom). The second half-filled orbital on the oxygen is bonded to the hydrogen (a third volunteer who entered with one pencil/electron). Ask the volunteers to simulate their respective electronegativities (the oxygen draws chlorine's bonding electrons closer to itself, indicating a stronger attraction for the electrons. It does likewise with the electrons bonding it to the hydrogen. You should not have to tell the students to do this, as they should understand the process based on their understanding of electronegativity. After all students have had a chance to observe and question, it is time to bring in an independent/non-bonded volunteer, who is the solvent water molecule, with a lone pair of electrons (dot structure of water if needed) which attracts the hydrogen atom and, to a small degree, ionizes the bond, creating a hydronium ion. Now we begin to make comparisons. Bring an additional oxygen atom volunteer into the system. This individual bonds to the lone pair of the chlorine atom. Ask the chlorine what he feels. "Anger? Why? Oh, someone is trying to steal your electrons? How does that make you feel with respect to electrical charge? Oh, electrons are negative, so you feel more positive? So how do you treat the electrons which you are sharing with the oxygen of the OH group?" Now to the hydrogen volunteer, "What are you feeling?"

"I'm more positive than before, and those water electrons are looking more enticing."

You, "And if more oxygen atoms attach, as in, chlorate or perchlorate?"

"I keep getting more positive and move to the water more readily."

After taking questions from students, you can now dismiss the original chlorine volunteer and bring in a new person. This individual can be a phosphorous atom, which has a smaller electronegativity than chlorine. "What does that mean?"

"It doesn't pull as hard on the electrons of the bonds." Have the atoms adjust their control of the electrons.

You, "So Mr. Oxygen," *who is bonded to the acidic hydrogen*, "how do you feel about this change?"

"I feel good. I get more control over those electrons. I'm not as positive as before."

You, "So what happens to the bond with hydrogen?"

"I have more of the central atom's electrons, so I don't need to pull on the hydrogen electrons as hard."

You, "And Ms. Hydrogen?"

"I'm not as charged, positive that is, so I am not as attracted to the water. I'm not as acidic."

You should also point out that, because most processes are reversible, the equilibrium point (*percent ionization*) will be affected by the tendency of the resultant anion to regain the proton. The more the negative charge is spread out, the less the tendency to regain the proton; i.e. it will remain ionized. Notice that you are sowing the seeds for equilibrium.

A similar sequence will occur as you complete successive ionization of a poly-protic acid. As each proton is removed, the electron density of all other bonds increases, reducing the polarity of the O-H bond and leading to a lesser attraction for the lone pair of electrons on the water and a weaker acid. Having the ability to feel this process adds credibility and logic for the students. I have observed students pause while answering a test question, hold their pencil with one hand then grip it with their second hand, pull, smile and return to writing. The process is effective.

You can now summarize: anything which makes the hydrogen atom more positive will make the acid stronger. To verify their understanding, ask which should be a stronger acid: acetic or trifluoro-acetic acid? Why, with explanation? As always, support the conclusion with percent ionization results. You may also address the question of solvent polarity and acidity, that is, what if we replace the water with methanol? Why?

Ammonia and amines are an obvious next step in the discussion. The idea that the O-H bond of water is now being severed by the nitrogen attraction for the water molecule's proton implies that the efficiency of this bond breakage will depend upon the relative charge of the nitrogen. Anything which causes it to become more negative will increase the attraction for the proton and yield a stronger base.

The next day, you can do a quick review by having Tom complete the ionization process for

$HCl + H_2O \rightarrow$

Then have Sally do the same for $HF + H_2O \rightarrow$
Then have Joe do likewise for $HOH + H_2O \rightarrow$

Whoa, isn't HOH water? Can it do that, can it ionize itself? If so, would it be very extensive? What evidence do you have? How could it be measured?

Note the range of hydronium ion concentrations as we go from a 1.0 M HCl solution to a 1.0 M HF solution to a pure water sample and to a 1.0 M NaOH solution. You are ready to introduce pH and the reason that we use the logarithmic scale. If you include equilibrium calculations in your introductory course, you are ready to define K_a, K_b and K_w. I typically saved the former two for AP but used the latter in general chemistry. Here again, my belief is that students should understand the principles and rationale before addressing the math. Many would see the math including equilibrium constants, square roots, etc. and shut down. The ideas of reversibility and equilibrium will remain long after the math has been forgotten.

I have a couple of other lessons to share. The first is a one-day exercise regarding the societal importance of some acids and bases. Near the end of a class period, I divide the class into six groups and charge each group with researching a particular acid or base and to be ready to share their findings. Each group member is to search for the information, as it is not always readily available. They are asked to find 1) the world financial ranking of the substance, 2) how much of the substance is produced annually worldwide, and in the US, and what it is worth, 3) what is the bulk price of the substance, 4) what are the common uses for the substance, 5) how is it made?

The assigned compounds are sulfuric acid, phosphoric acid, nitric acid, ammonia/ammonium hydroxide, sodium hydroxide and calcium hydroxide/calcium oxide.

The following day, the students enter the room where the six chemicals are listed on the board or separate overheads. Group members fill out the information, and then we share with the class as a whole. Sulfuric acid, once known as King Acid and used as an indicator of a nation's degree of industrialization, leads the way. It is used in many ways, including the production of several other acids on the list. Phosphates, nitrates and ammonia are all key materials in the agricultural world, serving as fertilizers, while lime and calcium oxide are used to adjust the pH of soil and to neutralize acids.

The exercise also gives you an opportunity to look at industrial processes such as pickling, leaching, saponification and soap production, the general methodology, the choice of reagent and resulting differences, hard vs. soft soaps for example, and other ways in which these substances affect our lives daily. Most students have no idea how ubiquitous these substances and the products made from them are in our lives.

The two demonstrations which I liked to include in this unit were side-by-side demonstrations. The first (see page 134) examines the constituents of a solution while it is being neutralized. The second centers on the sequential ionization/neutralization of a diprotic acid as well as the reaction of carbon dioxide with hydroxide ions (see page 135).

There are many standard acid/base experiments available including titrations, titrations using a pH meter, determining the acid content of vinegar, determining the equivalent weight of a solid acid, and so forth. Most are very good and worthwhile. I encourage you to have your students perform some of them **IN THE LABORATORY.**

Experiments and Demonstrations

"Chemistry is necessarily an experimental science: its conclusions are drawn from data, its principles supported by evidence from facts."
Michael Faraday

It would be great if we had unlimited student contact time and unlimited physical facilities, as nearly all chemistry could then be fleshed out in the laboratory. Unfortunately, we do not. So it is important that we use our laboratory time, discussion, and demonstration time wisely. In the pages that follow, I discuss several experiments and demonstrations which I found to be very useful and why I like them.

I often thought about the question, "Should this exercise be planned as a demonstration or as a class lab exercise?" Eventually, I reached a decision that it would be the latter if it taught a particular lab skill, such as the proper use of a Bunsen burner or burette, a technique such as vacuum filtration, reflux or distillation, or if it produced many sets of data which could be shared by the class as a whole such as the copper(II) sulfate and zinc lab (Appendix 1), if it required individual planning and preparation such as the two gram lab or qualitative analysis as described below, or if it required significant interpretation and analysis. Of course, the laboratory can also be used for evaluation of students.

Demonstrations were frequently unplanned and a response to a student question, such as, "Will all carbonates release carbon dioxide when exposed to excess acid?" We would randomly try a couple of carbonates and hydrogen carbonates and see that they all 'fizz.' Then we would examine why this occurred. Seeing and doing are better than being told, and an immediate visual response today is better than a staged demonstration tomorrow. Use your background and be open to this type of response. I realize, however, that not everyone has the supply room through a door in the rear of the classroom, and some questions are posed at a time which would interrupt the flow of ideas or prevent completion of material before the bell. Experience will guide your actions, but keep the spontaneous demonstration in your bag of effective teaching tools.

Many demonstrations were planned, of course, and were an integral part of the lesson. These include the decomposition of nitrogen triiodide and rapid combustion of peroxyacetone when discussing the instability of a weak covalent bond. Some were designed to challenge understanding or apply knowledge to a new situation. These included two demonstrations which I referred to as 'side by side' demonstrations. The first was inspired by an old *AP Chemistry Free Response* question[14]. The question states that a barium hydroxide solution is placed in a beaker and dilute sulfuric acid is dripped into the stirred solution. It goes on to describe the conductivity of the solution and requests an explanation of the observations. I found students readily picked up on the neutralization of the base or the precipitation of barium sulfate, but few picked up on both. So, I set up two similar systems both employing barium hydroxide as in the test question, but I placed sulfuric acid in one burette and hydrochloric acid in the other. With sulfuric acid, students observed the dimming of the light bulb, the change of phenolphthalein color and then the solution again becoming a strong electrolyte as more sulfuric acid was added. Then we repeated with the hydrochloric acid. The indicator again changed color, but the light never dimmed during the process. After observing, students were grouped into teams of three and asked to discuss and then record their interpretation of what they had observed. The group discussion invariably solidified both the neutralization and precipitation processes.

The second side by side demonstration was the result of a student question, "Does it matter which reactant is added to the other?" which led to a published article[15]. In the demonstration, two identical set-ups were prepared. Each contained a 50 ml burette inserted into a 2-hole stopper, which was affixed to a 125 ml Erlenmeyer flask. The second hole contained a glass tube with a hose for gas collection, a pneumatic trough and an inverted 1 L graduated cylinder to collect the carbon dioxide by water displacement. Then 50 ml of 1.0 M sodium carbonate was added to flask #1 and burette #2. 1.0 M hydrochloric acid was then placed in flask #2 and burette #1. 25 ml of solution were then slowly dripped into the flasks from the burettes as the solutions were stirred. At this point, we would pause and read the volume of gas collected in each cylinder. Then the remaining 25 ml of solution were slowly added, and the burettes were closed after reaching 50 ml added. Volumes of gas are again read. Students were then put into small groups and again asked to explain their observations. It was common that, after a few minutes, one of the groups would ask if they could add a little more acid to system #1 in order to check their hypothesis. The result always brought smiles and high-fives.

The second demonstration can also be used in AP Chemistry, but if this is done, the students should also be quizzed on why the second system finds water being 'siphoned' from the trough into the reaction flask after the final addition of carbonate is made. I did not expect first-year students to pick up on the reaction of carbon dioxide with the hydroxide ion to form the hydrogen carbonate ion, thereby reducing the pressure inside the flask. However, AP students should recognize this common reaction.

I used this process as a demonstration because, again, I wanted the students to discover the stoichiometry and the excess vs. limiting reactant implications. Writing 2 $HCl + Na_2CO_3 \rightarrow$ doesn't imply that is what is available in the reaction vessel. This demo could be used at various times throughout the year, but I found it most useful during the acid base unit.

I believe demonstrations can be a powerful addition to a course but should be used to solidify an idea or concept, not for their '*Wow!*' effect. In fact, I think the best demos bring a '*Huh?*' and not a '*Wow!*'

Additional Experiments

"If we knew what we were doing,
it would not be called research, would it?"
Albert Einstein

I have included sketches of several experiments within the units but also have some which do not fit into a given unit or could be used in various places or, as with the first case, are only done when the school calendar and Wisconsin weather allow time. Then we introduce the topic of acid media or base media redox reactions, and we run an experiment in which students titrate ferrous ions with permanganate ions to determine the concentration of a stock permanganate solution. The five-to-one mole ratio allows very precise and accurate results. This is a common experiment found in numerous lab manuals, but I give an application not available everywhere. Much of Wisconsin is blessed to have well water which is high in iron and calcium ions. This provides a nice business for water softener firms. Since many citizens own their own well, it also provides a source of unsoftened water. So I have students bring in a fresh sample of water from their well using a plastic water or soda bottle. After completing their stock solution titrations, they dilute the stock by 1:100 and then titrate three well water samples with the diluted permanganate solution. Results are not incredibly accurate but fall within the approximate known range of iron concentrations in the local well water.

Note: it is important that they run the water to clear the pipes and hose before filling the bottle and then fill completely, as oxygen will affect their results.

The second experiment became our culminating activity early in my career. I recalled doing qualitative analysis in high school and again in General College Chemistry. The procedure always struck me as 'cookbook' and was viewed by many as 'magic' with only a few chemistry majors appreciating the reactions and logic hidden in the subtleties of controlling pH, volumes, oxidation states and such. So I created a different approach. I told the students that they must create a test or series of tests to identify certain ions. They were free to use any safe available chemical. The first five anions they were to differ were carbonate, sulfate, sulfite, sulfide and phosphate. These are easily identified by their precipitates or gas formation with a dilute acid. Students recalled knowledge from throughout the year to develop a procedure. After they were satisfied that they could identify each, I gave them three 'unknown' salts which contained one of the five anions or an unreactive species (such as nitrate or acetate). After completing the unknowns, they were presented with five more anions (nitrate, nitrite, iodide, bromide and chloride). They needed to differ these from each other **and** from the first five. After developing tests to identify four of the five, I demonstrated the nitrate test accompanied by stern warnings that this was to be a confirming test after eliminating the other nine anions. Then they received three more unknowns containing any of the ten anions. After this, they proceed to three sets of cations and ultimately the identification of three salts where they had to provide the formula of the compound.

By referring to past notes and lab results such as those observed during the double replacement experiment (see Appendix 2), I found some students using precipitation reactions while other formed gases or completed oxidation/reduction reactions, and others employed single replacement reactions or physical tests such as flame tests. In later years, of course, some consulted the internet for procedures, but most accepted my challenge to use their knowledge and experience to develop pathways. After completing the lab work, they were assigned the task of creating two flow charts including net ionic reactions for chemical tests and a description of physical tests. *(What is the source of color in a flame test?)*

If you think this lab exercise sounds interesting, please proceed with open eyes. The first time I tried it, I was working at an open-campus school, and I had two students get in trouble because they were skipping English class to spend more time in the lab. I assumed they were coming from study hall or a free period and welcomed their efforts. Oops. Even after many years of using the procedure, I had trouble making students clean up in time to leave a neat laboratory. Also be aware that the appearance of some precipitates is dependent upon the identity or concentration of other ions present. This is particularly true of iron(II) and iron(III) products, which may vary from green to yellow or orange depending upon conditions. Thus, a student might test the stock iron(III) solution and receive a green-brown solid but test his unknown sample and receive a yellow-green product, causing him to reach the wrong conclusion. I learned to ask them how they reached the false conclusion. If they showed me their notes and the test result, I would suggest a test which would clearly separate the two ions in question, perhaps a thiocyanate ion test in this scenario. If they have developed a set of tests in good faith, I feel they should not be penalized for such variations. Incomplete or erroneous notes, however, did not warrant such assistance.

If you try this with your students, YOU will learn a lot of descriptive chemistry. For example, a solubility table will indicate that calcium sulfate is insoluble, but mixing dilute solutions of the two ions will not produce an immediate precipitate. If set aside for a while, the solid slowly forms. A few helpful hints if you proceed. First, have plenty of samples prepared in advance. After a few years, I purchased hundreds of sample vials with caps so that I could prepare fifty samples or more for each ion, using several complementary ions of low reactivity (example: sodium and potassium ions during anion analysis). This decreases identification solely based on appearance of the salt.[a] Having so many samples will also require that you have an easy method for determining if a reported ion is correct or not. I created a simple but not obvious code which told me the identity immediately. The code involved a letter followed by two numbers, another letter and a final number. If the first letter was A through F, the unknown was from the first set of anions. The unit digit of the sum of the two numbers gave the specific ion. For example, if the unit value was '1' the sample was phosphate. Thus, A10T6 and B38Z3 were both phosphates because both 0+1 and 3+8 have a unit value of '1' (1 and 11). However, T10B6 and C45P1 are not phosphates. As an aside, I always offered bonus points to those who could break part or all of the code. I told the students that nature has many hidden codes, and so as scientists it is our job to crack these codes. Take every opportunity to develop their minds. While challenging both for the student and the instructor, the method offers many positives. First, it fosters good observation skills, good note taking and a systematic approach. Second, it furthers the idea that there are multiple pathways to the solution of most problems. Third, it encourages multiple and creative solutions. Fourth, it offers an opportunity to review many notes without realizing it: a valuable activity as you approach final exams. It provides a solid foundation from which you can discuss selective precipitation, polyprotic acid equilibria, 'cookbook' qualitative analysis and many other topics in a later course. Very importantly, it furthers the curiosity factor and is a very enjoyable procedure for most students. When surveyed, this experiment and the two-gram lab, described elsewhere, are cited as the two favorite lab exercises by nearly all students. Not surprisingly, both are heavily student-driven, and success was attainable with a bit of insight and effort.

At the end of this unit or during AP Chemistry, you can challenge and reinforce learning by developing special problems. For example, some years ago I shared an identification problem with the writers of the *Chemistry Olympiad Exam Part C*[16]. Students were given samples marked A through F containing equal concentration solutions of hydrogen carbonate, carbonate, hydrogen sulfite, sulfite, phosphate and hydrogen phosphate, along with a similar concentration solution of hydrochloric acid and phenolphthalein. They were asked to identify each solution. The analysis required both qualitative and quantitative work, including counting drops and watching for bubbles and/or checking for odors: a challenging but very solvable problem.

Students often ask for additional lab experience to reinforce their understanding. Encourage this as much as possible. These exercises may be a full-blown investigation, or they may be just a 'quick-and-dirty' set of tests. For example, a student asked about IMFs and macroscopic properties like boiling points and evaporation. I sketched the structure for acetone, hexane, water, glycerine and ethanol, then asked her to rank them from weakest IMFs to strongest. After doing so, I gave her five identical beakers and instructed her to carefully mass each and then add 20 ml pure liquid to each, mass the beaker and contents, then re-mass several times during the day and the next morning before school. She was then to determine the mass of liquid remaining and graph the results. Later, she presented the study to her classmates, properly explaining not only the sequence of evaporation but also the reason for the increasing mass by glycerine. A study need not be fancy to be insightful. A similar quick and effective study can be done during the qualitative analysis work. If a student finishes early, present him with a sample containing a viable salt which has been thoroughly mixed with a nonreactive substance (example: barium chloride with potassium acetate). He must not only identify the salt but also determine the percent by mass in the sample. Similarly, a more complex study can be completed. One I offered was to use pure calcium carbonate with hydrochloric acid to generate data which could be used to determine the purity of calcium carbonate in limestone. What assumptions are being made?

It is the experience of most students that day one in a class consists of dos and don'ts, grading criteria, and a syllabus. Using my 'Golden Rule' criteria, I realized this is incredibly boring, so I adopted a 'you can read' approach, highlighting only key topics such as my honesty policy, and used at least half of the day to take them to the lab where they observed or completed several tests. Of course, they have not yet signed a Safety Contract at this point, so be careful selecting tests. They were told that I would collect their explanation of each test and count it as part of their semester exam. All of the phenomena or reactions were discussed during that upcoming semester. I called this the '90-day assignment.' Be certain that you remind them that their answers are due at your selected date prior to the exam.

AP EXPERIMENTS AND OTHER COMMENTS

You have probably noticed that I have mentioned AP Chem several times without much discussion. That was because I found the curriculum and lab expectations of the course to be well defined and explicit: there was little room or time for digression. Occasionally, however, I saw what I considered to be a flaw and sought to correct it. One of these was during the bonding unit. I believe Molecular Orbital Theory is too important to ignore in a college introductory level. As a result, I would tell the students to tune in but not worry about the subject appearing on any exam. Then we would explore the general ideas behind MO Theory (all of my students would have studied waves in physics by this time) and then look at bonding between two atoms, paying particular attention to the magnetic properties of oxygen. Nothing too heavy, but at least an introduction which can churn in the back of their mind until a more in-depth study is made.

A second thing which bothered me was that none of the AP Lab Manuals seemed to provide an experiment focused on thermodynamic properties. There are several nice studies dealing with kinetics (fading of phenolphthalein, iodination of acetone, etc.) and equilibrium, including Le Chatelier's Principle and determination of equilibrium constants, but nothing for thermodynamics. Consequently, I challenged two seniors who had completed AP Chem as juniors to carry out some research. The result was a nice experiment using *Vernier Labware* to determine the enthalpy change, entropy change, free energy and equilibrium constant for a reaction commonly used to demonstrate Le Chatelier's Principle and a publication for the young ladies, one of whom pursued physics and the other chemistry.[17]

While on the subject of AP, I want to address one other topic. For many years, I used the post-AP-exam time to provide additional lab experience. After a few weeks of lab work without mandatory reports, coupled with approaching graduation for most students, however, senioritis set in. Another option was needed. With the AP Physics teacher, Ms. Cheri Kaiser, I established a series of presentations by alumni who were in graduate school or working as engineers, scientists or physicians. Alumni response was incredible, as they were eager to give back to the school and current students. After several years, graduates began contacting us to see if we would host them as they grew in their field. Their presentations included topics such as carbon nanotube technology from a chemical engineer with several patents in the field, physicians who performed skin grafts on burn and cancer patients, architectural engineers, automobile safety engineers, food chemists, string theory physicists, astronomers, and many others. These seminars were well-received and led several students to pursue a field which was previously unknown to them. Obviously, you cannot develop such a program in your first year, but keep it in mind for later. It is a great opportunity for you to learn and provides good public relations for the school.

[a]Inexpensive vials with caps may be purchased from American Science and Surplus. You may request their catalog, an eccentric and creative work in its own right.

Appendix 1

Job's method applied to heat of reaction

INTRODUCTION

This reaction involves a single replacement process. Both reactants and products involve soluble and insoluble materials. However, the release of energy in the form of heat allows us to monitor the process. We will use the temperature change and the colors of solution and residue to determine the best possible mixture for the reactants.

PROCEDURE

Each lab group should be assigned a reaction mixture from those below. The source of copper(II) sulfate is a 1.00 M solution which I have prepared. The zinc is in powdered form.

You should calculate the quantity of solution needed, then carefully measure that amount using an appropriate size graduated cylinder. This solution should be carefully diluted with water to achieve a total of 100 ml. Carefully mass the quantity of zinc needed for your experiment.

Carefully pour the copper solution into a double-nested Styrofoam calorimeter which is in turn nested in a beaker. Insert a thermistor and record the temperature. Add the zinc and stir thoroughly while monitoring the temperature.

When the reaction is complete, decant the liquid into a beaker and pour the solid onto a piece of white paper supported by paper toweling. Label each with the station number and arrange numerically. Carefully observe and record the various results. Share the data by entering your results in the computer program "Zn Lab" provided. A sample of the data requested is given below.

GROUP ASSIGNMENTS

Group #	Moles Zinc	Moles of $CuSO_4$ sol'n	Milliliters of $CuSO_4$ sol'n	^T
1	.09	.01	10	
2	.08	.02	20	
3	.07	.03	30	
4	.06	.04	40	
5	.05	.05	50	
6	.04	.06	60	
7	.03	.07	70	
8	.02	.08	80	
9	.01	.09	90	

Appendix 2

Double Replacement lab   Name _____

INTRODUCTION

When double replacement reactions occur, any of several results may be observed. During this exercise, you will be asked to carefully observe a number of mixtures. Carefully note and record any observations [formation of a gas (odor?), heat, solid formation, etc.]. You will be asked to determine which **PRODUCT** was responsible for the observed change.

PROCEDURE

Starting at your lab station, use a clean dropper to add a few drops of reagent #1. Then use a clean dropper to a few drops of reagent #2. Observe, record, clean the test tube and proceed to the next pair of reactants. When completed, balance each reaction, and indicate the product responsible for the observed change.

1. $KNO_3 + NaCl \rightarrow$
2. $K_2SO_4 + BaCl_2 \rightarrow$

3. $KOH + H_2SO_4 \rightarrow$
4. $Na_3PO_4 + CoCl_2 \rightarrow$
5. $HCl + NaHSO_3 \rightarrow$

(Stations 4, 5, & 6 are set up in the hood)

6. $(NH_4)_2S + HCl \rightarrow$
7. $H_2SO_4 + KOH \rightarrow$

8. $FeCl_3 + (NH_4)_2S \rightarrow$
9. $HCl + NaHCO_3 \rightarrow$

10. $NaOH + HCl \rightarrow$
11. $BaCl_2 + CuSO_4 \rightarrow$

12. $NH_4Cl + NaOH \rightarrow$
13. $AlCl_3 + NaOH \rightarrow$
14. $K_2SO_4 + MgCl_2 \rightarrow$
15. $Pb(NO_3)_2 + KI \rightarrow$

16. $NaBr + AgNO_3 \rightarrow$
17. $K_2CO_3 + HCl \rightarrow$
18. $Ni(NO_3)_2 + KCl \rightarrow$

Appendix 3

Solute	# g	# moles	FP
CH_3OH	32	1	-1.86
CH_3OH	64	2	-3.72
CH_3OH	18	0.5	-0.93
CH_3OH	8	0.25	-0.46
C_2H_5OH	46	1	-1.86
C_2H_5OH	92	2	-3.72
C_2H_5OH	138	3	-5.58

Solute	# g	# moles	FP
$C_2H_6O_2$	62	1	-1.86
$C_3H_8O_3$	92	1	-1.86
$C_6H_{12}O_6$	180	1	-1.86
$C_{12}H_{22}O_{11}$	342	1	-1.86
As_2S_3	244	1	-1.86
NCl_3	119	1	-1.86
PCl_3	136	1	-1.86

Solute	# g	# moles	FP
NaCl	58	1	-3.72
NaCl	29	0.5	-1.86
Na_2SO_4	142	1	-5.58
$NaNO_3$	85	1	-3.72
$NiSO_4$	154	1	-3.72
AlF_3	84	1	-7.44
$BaBr_2$	297	1	-5.58

Solute	# g	# moles	FP
KBr	119	1	-3.72
KBr	60	0.5	-1.86
$CuSO_4$	159	1	-3.72
$AlPO_4$	122	1	-3.72
$NiCl_2$	129	1	-5.58
Al_2S_3	150	1	-9.3
ZnI_2	319	1	-5.58

Solute	# g	# moles	FP
HCl	36.5	1	-3.72
HNO_3	63	1	-3.72
H_2SO_4	98	1	-5.58
HF	20	1	-2.23
HF	40	2	-4.46
$HC_2H_3O_2$	60	1	-1.88
H_2SO_3	82	1	-2.1
HBr	81	1	-3.72
H_2CO_3	62	1	-2.01
$HClO_4$	100	1	-3.72
HClO	52	1	-1.95

Appendix 4

% Ionization of 1.0 M Acids

Acid	Percent Ionization
HCl	100
HBr	100
HI	100
HNO_3	100
H_2SO_4	110
$HClO_4$	100
HF	2.68
$HC_2H_3O_2$	0.42
H_2SO_3	10.9
HNO_2	2.1
H_3PO_4	8.7
$H_2PO_4^{1-}$	0.028
HPO_4^{2-}	7.9E-05
HSO_3^{1-}	0.026
$HClO_3$	100
$HClO_2$	10.5
HClO	1.73

CITATIONS

[1] Rhodes, Richard. *The Making of the Atomic Bomb*, New York, New York: Simon & Schuster, 1986.
[2] Zipp, Arden P. *Journal of Chemical Education*, 1992, 69, 291.
[3] Graham, D.M. *Journal of Chemical Education*, 1989, 66, 1989, 573.
[4] Diemente, D. *Journal of Chemical Education*, 1998, 75, 1565.
[5] Johns, P.T. *Journal of Chemical Education*, 1993, 70, 774.
[6] van Lubeck, H. *Journal of Chemical Education*, 1989, 66, 762.
[7] Hawthorne Jr, R.M. *Journal of Chemical Education*, 1973, 50, 282
[8] Antony, E. *Journal of Chemical Education*, 1991, 68, 1040-1041.
[9] Johnson, George (2009) *The 10 Most Beautiful Experiments*, New York, New York: Random House.
[10] Reed, James L. *Journal of Chemical Education*, 1999, 76, 802.
[11] Shakhashiri, Bassam Z. (1983) *Chemical Demonstrations*, Madison, WI, University of Wisconsin Press.
[12] Zumdahl, Susan A, Zumdahl, Steven S. (2007) *Chemistry*, Boston, MA, Houghton, Mifflin, Harcourt.
[13] Hawkes, Stephan J. *Journal of Chemical Education*, 1998, 75, 1179-1181.
[14] AP Chemistry Exam, Free Response #7, 1982.

[15] Antony, E.; Mitchell, L.; Nettenstrom, *Journal of Chemical Education*, 2000, 79,1180-1181.
[16] National Chemistry Olympiad, Part C 2009.
[17] Antony, E.; Muccianti, C.; Vogel, T.; *Journal of Chemical Education*, 2012, 89, 533-535.

Professionalism/Personal History

"An expert is a person who has made all the mistakes that can be made in a very narrow field."
Niels Bohr

There are many levels of professionalism: professionalism in the classroom, professionalism in the building and beyond. I will leave it to the schools of education to address the components of the classroom and building levels of professionalism. I believe a common thread throughout the various levels of professionalism is your personal attitude and your beliefs. Do you teach just so that you can coach? Do you teach for the growth and success of your students? Do you teach for the growth and success of all students? I would encourage those in the first group to seek a new career. I applaud those in the second group but would encourage them to expand their visions to become part of the third group as well. Doing so could include things like subscribing to, reading, and sharing articles from journals, especially the *Journal of Chemical Education*. You probably won't have time to do so during the school week, but you can catch up over a weekend or vacation, sharing pertinent and 'of interest' articles with peers and, when appropriate, students.

Present ideas or methods at local, state and national conventions. Publish new and worthwhile lessons and experiments or demonstrations, become active with the *American Chemical Society* or state science teacher organizations by attending meetings and assisting with programs like the *Chemistry Olympiad* or *JETS Testing*. When possible, include peers and students in these activities. Most of my publications have included peers or students as coauthors.

These actions may not be possible or practical in the first years of teaching as you are learning the art and balancing time demands. But as you gain experience and maturity, you should strive to become active beyond the classroom and building. If your contributions help a student in a neighboring school become a chemist and that student discovers the cure for Alzheimer's disease, it may not be a feather in your hat, but it is for mankind. Do it.

During my early years of teaching, I continued the college habits of reading some chemical literature. These included *Chemical and Engineering News* the weekly publication of the *ACS, Science Teacher* a publication of the *National Science Teachers Association, Reviews of Chemical Research* and the *Journal of Inorganic Chemistry*, my personal area of greatest interest. These primarily satisfied my personal interest, having little direct effect upon my classroom. During the first few years, I also carried out some research on synthesis of cyclic amine metal ion complexes. Since this was strictly for personal interest, I purchased both the amines and the metal salts with personal funds as I had done with the periodicals.

As noted in the introduction, I have few comments regarding district level work. That should not suggest that such work is not meaningful. Over the years, I worked with peers to develop a course on the subject of color. The course included the effect of colors on our psychology taught by the AP Psychology teacher, the effect of color on our moods in interior decorations and appetites to be taught by a FACE instructor, color in art, as well as the chemistry and physics of color by reflection, refraction and absorption. We also developed brain-based education units using a peer research and review approach, including outstanding educators from fields including art, English, counseling and the sciences. Such collaborations can bring uniformity to our efforts and approach, thereby enhancing the students' learning and understanding. Do not neglect such opportunities.

My education-based professional activities outside the district began about my eighth year of teaching. Wisconsin has a statewide two-day teacher convention each fall. I soon learned that most sessions dealt with elementary school, middle school, Special Education or reading initiative programs. I sought and received permission from local administrators to host a two-day conference for area chemistry teachers. We met and shared thoughts, experiments and materials in a very worthwhile manner. This soon led to presentations of my own at conventions. Shortly thereafter, I secured a position as department chair at an eastern regional magnet school. This position put me in contact with many business leaders and chemists, which then led to my being active as a distributor of used instruments and lab equipment: businesses, universities and medical laboratories donated volumetric flasks and other glassware, IR Spectrometers, gas chromatographs and spectrometers which were then donated to local school districts.

I continued this program shortly after returning to Wisconsin, where I joined the Chemical Education Committee of the Milwaukee Section of the ACS, and a few years later I assumed leadership of the committee. This was a position which I held over ten years. Here we funneled used materials and chemicals to local schools. One received a used infrared spectrophotometer to augment their organic chemistry course, while others gained glassware, UV/visible spectrophotometers and such. One local university even received a used NMR from a firm in New York City. Given the shrinking budgets of educational institutions, these donations were all appreciated. An area medical lab closed and donated chemicals, including ten bottles of silver nitrate (which were quickly claimed by schools throughout the section). Another donated several thousand unused one-liter plastic bottles. Again, this was work which helped students throughout the area. Caution, however. Before blindly accepting chemical contributions, be certain they are safe and that you or others need and will accept the chemicals, or you may end up paying for the removal and disposal of the substances.

During this time, I helped grade the AP Chemistry Exam for several years before a change in our school calendar prevented my continuation. This again is a good, professional activity for those involved in the AP program. The camaraderie is great, and evenings are spent discussing experiments, activities, and philosophy with some of the best teachers in the country and beyond. The AP program regularly posts requests for more readers. I was also fortunate to secure a position as grader of the National Chemistry Olympiad exam for several years. The position claimed a weekend every spring without pay but allowed me to interact with individuals such as Dr. Thomas Holme, Mr. Steve Lantos, Dr. Ronald Ragsdale and particularly Dr. Arden Zipp, Cortland College: individuals who had long and distinguished careers in chemical education. The discussions with these individuals helped shape my lab design and lab-based evaluations over the ensuing years. I continued with the group until the ACS Exam Institute was moved from Milwaukee to Ames, Iowa. I was also honored at one point by being the first high school teacher to be invited to join the Chemist Circle of Milwaukee. After several meetings, I found participation to not meet my personal interests and needs, so I resigned from the group.

I understand that all instructors cannot be as involved as I was while some are more active, taking part as trainers for the National Chemistry Olympiad team, the Woodrow Wilson Institute, assisting with the Institute of Chemical Education program or other summer training programs. The point is, push yourself to be as active and professional as possible.

www.ingramcontent.com/pod-product-compliance
Lightning Source LLC
Chambersburg PA
CBHW071222090426
42736CB00014B/2933